Great Ideas of Classical Physics

Part II

Professor Steven Pollock

THE TEACHING COMPANY ®

PUBLISHED BY:

THE TEACHING COMPANY
4151 Lafayette Center Drive, Suite 100
Chantilly, Virginia 20151-1232
1-800-TEACH-12
Fax—703-378-3819
www.teach12.com

Printed in the United States of America

ISBN 1-59803-255-0

Steven Pollock, Ph.D.

Associate Professor of Physics, University of Colorado, Boulder

Steven Pollock is associate professor of physics at the University of Colorado, Boulder. He did his undergraduate work at MIT, receiving a B.Sc. in physics in 1982. He holds a master's and a Ph.D. in physics from Stanford University, where he completed a thesis on "Electroweak Interactions in the Nuclear Domain" in 1987. He did postdoctoral research at NIKHEF (the National Institute for Nuclear and High Energy Physics) in Amsterdam from 1988–1990 and at the Institute for Nuclear Theory in Seattle from 1990–1992. He spent a year as senior researcher at NIKHEF in 1993 before moving to Boulder.

From 1993–2000, Professor Pollock's research work focused on the intersections of nuclear and particle physics, with special focus on parity violation, neutrino physics, and virtual strangeness content of ordinary matter. Around the time he received tenure at CU Boulder, Professor Pollock began shifting his attention to the newly developing discipline-based research field of physics education research. This field now represents his full-time physics research activities.

Professor Pollock was a teaching assistant and tutor for undergraduates throughout his years as both an undergraduate and graduate student. As a college professor, he has taught a wide variety of university courses at all levels, from introductory physics to advanced nuclear and particle physics, including quantum physics (both introductory and senior level) and mathematical physics, with intriguing recent forays into the physics of energy and the environment and the physics of sound and music.

Professor Pollock is the author of *Thinkwell's Physics I*, a CD-based introductory physics "next-generation" multimedia textbook. He became a Pew/Carnegie National Teaching Scholar in 2001 and is currently pursuing classroom research into replication and sustainability of reformed teaching techniques in (very) large lecture introductory courses. Professor Pollock received an Alfred P. Sloan Research Fellowship in 1994, the Boulder Faculty Assembly (CU campus-wide) Teaching Excellence Award in 1998, and the Marinus G. Smith Recognition Award in 2006. He has presented both nuclear physics research and his scholarship on teaching at numerous conferences, seminars, and colloquia. He is a member of the American Physical Society, the Forum on Education, and the American Association of Physics Teachers.

Acknowledgments:

Many thanks to David Steussy and Charles (Max) Brown for their assistance in creating ancillary materials for this course!

Table of Contents
Great Ideas of Classical Physics
Part II

Great Ideas of Classical Physics

Scope:

Physics is the science that tries to understand the deep principles underlying the world we live in. It's about understanding and describing nature. It's about *things*, as opposed to biological or even chemical *systems*. *How* do things move? *Why* do they move? How do they *work*? Physicists search for deep patterns, for the fundamental simplicity and unity of measurable phenomena. In this course, we will follow a theme-based, quasi-historical path, highlighting the central concepts, ideas, and discoveries of classical physics. *Classical* here refers to scientific work done up to the start of the 20th century, that is, essentially all physics before the quantum theory and relativity. It is the physics of everyday life, the physics of a deterministic "clockwork" universe, with enormous explanatory and predictive power! We will spend a little time getting to know the characters who played key roles, including Galileo, Newton, Faraday, Maxwell, and others, but the emphasis of the course is on sense-making: What have physicists learned about the world? What are the key underlying laws of nature? What are the primary organizing principles? How can we use these ideas and connect them to our personal experiences?

Physics is a broad field of study and can be approached from many angles. We begin with a venerable branch of physics known as *mechanics*, the study of forces, energy, and motion. The word *mechanics* might make one think of car engines, and in some ways, that's a good metaphor. Engines are complicated, but they are built out of simple and comprehensible parts, each of which serves a simple purpose. When put together, they create a familiar, useful, and understandable (by mechanics!) whole. But *mechanics* in physics is not about cars; it's the study of how just about anything moves and what makes objects behave as they do when acted on by forces. It's a study that will help us understand a vast and disparate array of phenomena, from Olympic high divers, to the display of sparks in a firework on the 4th of July, to the path of the Moon in the night sky, or the ceaseless bounce and jitter of atoms in a gas. We will focus on the central concepts: What do we know, and how do we know it? We'll ask where the ideas came from and how we might test them. And, of course, we'll ask what we can do with this knowledge. Classical mechanics is primarily the physics of Isaac Newton and a host of other brilliant characters who laid the groundwork for understanding the world that is still relevant 400 years after its beginnings. Our goal is to walk away with a sense of the order and coherence, the basic structure and principles of this foundation of physics.

Mechanics sits underneath the rest of physics a bit like the foundation of a great cathedral. The second half of the course will add the edifice, structure, and turrets. We will need to understand the ideas behind *electricity* and *magnetism*, forces that dominate our technological world and lead to understanding of the structure of all matter and light. This investigation leads naturally to *optics*, which was unified with electricity and magnetism in a brilliant stroke in the mid-1800s. In this context, we will briefly consider *waves* and the myriad phenomena that become understandable, and intimately related to one another, once we grasp the basic ideas and consequences of vibrations. We will need to learn separately about *heat* and *thermodynamics*, a branch of classical physics that deals with everything from understanding car engines and power supplies to making a perfect cake. This course of study takes us right up to the start of the 20th century.

One final comment: Mathematics plays a special role in science, one very dear to physicists, but we will not (and need not) focus on math in this course. Although skipping the equations limits, to some extent, the depth to which we can learn physics, the concepts themselves are, by and large, sensible, intuitive, and comprehensible through metaphor, life experience, ordinary logic, and common sense. From time to time, however, we may follow brief mathematical detours to appreciate the power and beauty of more formal or symbolic reasoning!

Notes on Course Materials: Suggested readings and computer simulations are listed with each lecture, using the abbreviations noted below.

Essential Computer Simulations ("Sims"):

These are all available at http://phet.colorado.edu and should run on PC or Mac. (Some of the Java applications require a fairly current Mac OS.)

Essential Reading:

Thinkwell	Professor Pollock's *Thinkwell Physics I*, www.thinkwell.com.
Hewitt	Paul G. Hewitt, *Conceptual Physics*, Addison Wesley, 2005.
Hobson	Art Hobson, *Physics: Concepts and Connections*, Prentice Hall, 2006.
March	Robert March, *Physics for Poets*, McGraw-Hill, 2002.

Recommended Reading:

Feynman	Feynman, Leighton, and Sands, *The Feynman Lectures on Physics*, Addison Wesley, 1963.
Cropper	William H. Cropper, *Great Physicists*, Oxford University Press, 2001.
Gleick	James Gleick, *Isaac Newton*, Vintage, 2003.
Lightman	Alan Lightman, *Great Ideas in Physics*, McGraw Hill, 2000.
Crease	Robert P. Crease, *The Prism and the Pendulum*, Random House, 2003.
Gonick	Larry Gonick and Art Huffman, *The Cartoon Guide to Physics*, Collins, 2005.

Lecture Thirteen
Further Developments—Static Electricity

Science is built up of facts, as a house is with stones. But a collection of facts is no more a science than a heap of stones is a house.
—Henri Poincare

Scope:

For 200 years following the publication of the *Principia*, growing numbers of scientists followed the path laid out by Newton—a path paved from a philosophical, mathematical, theoretical, and experimental groundwork. The scope of "physics" expanded steadily and rapidly, and we can only touch on the many "great ideas" developed in this period: electricity, magnetism, waves, optics, and the grand unification of those ideas. Heat and temperature, chemistry, and the atomic worldview make up another path for us to follow. We will study some of these grand ideas in upcoming lectures—a few very briefly and some in more detail—to get a sense of the sweeping scale of accomplishments of classical physics. In this lecture, we begin our story of post-Newtonian classical physics with the "new" forces of static electricity and magnetism. We'll look at static electricity as a classic example of a systematic investigation into a force of nature, but we'll see that this "new" force still fits in tightly with the Newtonian framework.

Outline

I. Let's begin with a road map for the rest of the course.

- **A.** In covering some new topics, we will always begin with Newton's ideas about forces, momentum, and energy and conservation laws.
- **B.** In the second half of the course, we will talk about the fundamental constituents of the world, particularly atoms and their motion. We'll see that the motion of particles is connected to theories of electricity and magnetism, as well as theories of light and optics. We will also explore thermodynamics.
- **C.** We will discover a new hero in this part of this course, James Clerk Maxwell (1831-1879), who is to electricity and magnetism what Newton was to the fundamental, underlying

laws of mechanics. As we'll see, electricity and magnetism are evident everywhere in our world, especially in our technology but also in basic structures.

II. In Newton's day, electricity was a curiosity. People were aware of the phenomenon of static electricity but didn't begin to investigate it scientifically until about 100 years after Newton.

A. We're all familiar with static electricity. Think of walking across a carpet, touching the doorknob, and getting a shock or pulling apart clothes that have just come out of the dryer. Whenever anything sticks together, like the clothes, that implies a force of nature, and in this case, the force is not just friction.

B. Let's investigate static electricity using a simplified approach.

1. Benjamin Franklin (1706–1790) helped start us on the path toward our current model of electric charges. In addition to flying the kite in the electrical storm, Franklin conducted experiments in which he rubbed various objects together, such as cat fur on amber or glass rods.

2. Try a similar experiment on your own: Take about a foot of tape and fold over the ends to make tabs for pulling the tape up. Label this piece of tape *b* for "bottom." Place a second piece of tape on top of the first and label it *t* for "top"; then, stick both pieces of tape, now stuck together, down on a flat, clean table. Next, duplicate the experimental setup. Play around with the tape by ripping it off the table, then ripping the two pieces apart.

3. It will be immediately obvious that the pieces of tape are charged. You'll also discover that different things happen to the top tape and the bottom tape. Two top tapes, for example, will repel each other, but a top and bottom will attract.

C. Let's construct a simple model to help us describe and understand the basic phenomenology of static electricity.

1. Recall our discussion of what a model is: a simplified, descriptive picture. It must be consistent with known experiments and lead us to predictions about future experiments.

2. In the accepted model of electricity and magnetism, the world is made of atoms that carry electric charge. Electric charge is the quantity that exhibits the force of static electricity.
3. The tape experiment shows us that we need two types of charge to explain the results. Ben Franklin named these types of charges *positive* and *negative.*
4. How did Franklin know that there wasn't a third type of charge? Ockham's razor (named for a medieval philosopher, William of Ockham) suggests that we use the simplest explanation to understand complex phenomena. In other words, we don't add another charge because we aren't required to by the data.

D. Franklin's choice of *positive* and *negative* for the names of the two charges is helpful to our understanding of electricity, because it leads us to think of adding *plus* and *minus* charges. When we add positive and negative charges, the net charge is zero.

E. According to our model, the world is filled with positive and negative charges, and as we deduce from the tape experiment, opposite charges attract and like charges repel.

F. Newton saw that the gravitational force between two masses arises because of the mass. In the same way, the electrical force arises because of the charge. As we'll see in an upcoming lecture, we find the strength of the electrical force by multiplying the charges. Again, with multiplying, the positive and negative sign convention neatly summarizes the fact that opposite charges attract and like charges repel.

G. Our model is predictive and explanatory, but it doesn't tell us what charge is; it only postulates that charge exists.

1. With our model, you can see that combing your hair separates charges; the comb becomes negatively charged and your hair becomes positively charged. Your hair stands on end because all those like charges are repelling one another.
2. You can also understand why a balloon will stick to you after you rub it on your shirt, but why does it stick to the wall? The answer is that the wall also has electrical charges in it. Any negative charges in the wall that are

free to move will be repelled by the balloon; any positive charges in the wall that are free to move will be attracted to the balloon. Separation of charges takes place again.

H. In one respect, Franklin's naming convention may be slightly confusing: We now use the term *electrons* for particles that carry negative charges, and these are the particles that move most easily in nature, although we might tend to associate positive charge with movement. Nonetheless, the simple story of static electricity has been spectacularly useful

III. Let's turn briefly to magnetism.

A. Find a child's set of magnets and experiment with them on your own. You'll find that magnetic force is quite similar to electric force. Instead of positive and negative charges, we say that magnets have *north* and *south poles*, which attract and repel each other analogously to charges.

B. One difference between magnets and static electricity is that magnetism seems to be permanent, whereas static electricity tends to fade with time for ordinary objects.

IV. We now have only a very basic understanding of static electricity. In future lectures, we'll see that the lightning bolt that Franklin was investigating; forces of nature, such as friction; and the high technology we now use all arise simply from the postulation of positive and negative charges and the electrical forces (ultimately, using Newton's law) between them.

Essential Computer Sim:

Go to http://phet.colorado.edu and play with Balloons and Static Electricity and John Travoltage. Do the balloons behave realistically? Are they conductors or insulators? How about the walls? Is charge flowing in the walls? Can you understand why the balloons stick to the wall, even though the total charge of the wall is zero? Why doesn't John Travoltage generate a spark immediately? Why does he have to build up some charge first? What role does the doorknob play? Why do you need it?

Essential Reading: Hewitt, start of ch 21, Hobson, ch 8.4, March, start of ch 6.

Recommended Reading: Gonick, chapter 12.

Questions to Consider:

1. Suppose somebody proposed to you that we are attracted to the Earth, not because of gravity, but because of static electrical forces. How could you convince this person that he or she is incorrect?
2. Try the tape experiment described in the lecture. What variations can you come up with? Can you determine which piece is positive and which is negative? Can you explain all your observations by hypothesizing only two charges (positive and negative), or do you need more?
3. Buy some toy magnets—not kitchen magnets but small bar magnets with two clear poles, like the plastic-coated, super-strong magnets that come with ball bearings as in a child's building kit. How are these magnets the same as and how are they different from the charged tapes? Can you prove that they are not attracting and repelling because of forces of static electricity? Can you build a compass out of these magnets?
4. Suppose that Ben Franklin had reversed his choice of which material to call *plus* and which to call *minus*. Would this change any of our laws of physics? What would be different?

Lecture Thirteen—Transcript
Further Developments—Static Electricity

Isaac Newton started something very big, and it wasn't just F=ma or the law of gravity. It was really a way of thinking about doing science. He was starting a new field of investigation. It was partly philosophical. It was partly that he was using mathematics to so powerful an effect. It was partly that he was constantly grounding what he was doing with experiments. These new ideas were always connected with direct experimental verification. It was really a frame for how you do science, how you investigate the world. For 200 years at least, in fact continuing today, after Isaac Newton and after the publication of the *Principia*, a growing number of scientists—and you can definitely call them scientists at this point—follow this path and the field of classical physics is growing. It's certainly not defined just by the content that Isaac Newton was investigating. We will go far beyond that content. It's really the task now for the rest of the course to try to lay some sort of path in the ever-expanding classical physics domain that was being developed. We're going to try to find the minimal subset of ideas and concepts to follow, which try to span the space of classical physics. It's not entirely possible, and there are some compromises you have to make. There will always be some topics of classical physics that we won't be able to discuss, but in the end we're going to cover those core ideas that allow us to explain the broadest array of physical phenomena and measurable quantities in the world.

Let's lay out a little road map for where we're headed. I want to start with Newton's ideas all of the time. Every time we're talking about some new concept, in the back of our mind we're thinking about forces, momentum, and energy in particular. These conservation laws and the laws of motion will allow us to make sense of anything that we want to investigate. If we want to study, for instance, electricity and magnetism, which are really important forces in our lives, we will be talking about new ideas, very new ideas, that Isaac Newton would have found quite alien, and yet at the same time they're grounded in how objects move and why they move that way.

We will also be talking about the fundamental constituents of the world. We will want to think about atoms and the motion not just of atoms, but the motion of any particle that's vibrating or wiggling. We'll see that this is connected to the theories of electricity and magnetism and also connected to theories of light and optics, and so in the end we'll try to put this all together. We'll look also at thermodynamics, which would be the explanation of and understanding of heat and temperature. In the end, when we've put all of that together, we will really have a frame for the kinds of questions that people continue to investigate today and which were studied and understood over this period of time following Isaac Newton when classical physics was in its real hay day.

We'll discover a new hero in this part of the course, James Clerk Maxwell. Maxwell and his equations define the theories of electricity and magnetism for us. Today, he is to electricity and magnetism what Isaac Newton was to the fundamental, underlying laws of mechanics. That's going to be the first topic that we'll want to build up to.

We'll discover that electricity and magnetism are everywhere in life. When you think about our technological lives, you can see the role of electricity in particular. Magnetism is a little bit more hidden, but we'll talk about it because you'll discover that magnetism is also everywhere in the technology that we use, just not quite as obvious or visible as the electrical phenomenon that we're studying. It's not just about the technology in the world we live in, but it's also about the structure of the world we live in. When you look at material objects and how hard they are or what color they are, or if you look at the phenomenon of light, you will understand the connections between electricity and all of these other aspects of the world we live in.

Back in Newton's day, electricity and magnetism were little curiosities. He was surely aware of the fact that in the wintertime when he brushed or combed his hair it might stand on its end or crackle a little bit. This static electrical phenomenon has always been observed, but people didn't really pay much attention to it. It seemed to be a little laboratory curiosity or some sort of funny thing that people didn't really understand. This obscure little force of nature began to attract people's attention, and in the 100 years post-Newton more and more scientists were trying to figure out this story. What is going on? It seems to be a new force of nature. It's very different from gravity although it has a few commonalities, and as we begin to develop our understanding of this phenomenon, historically people realize that yes indeed we can investigate this in the way that Isaac

Newton would have. We can investigate it making use of his concepts of force and acceleration, momentum conservation and energy conservation.

For today's lecture, we're going to start with static electricity. Static electricity is just what you think it is. You walk across the carpet, you build up static charge, you touch the doorknob, and you experience a shock. We like to understand what the physical underpinning phenomena are here. You take the clothes out of the drier, you pull them apart, they crackle, and they stick together. Anything that sticks together implies a force of nature, and you could lump it together with friction, but it becomes obvious pretty quickly that there's something different about static electricity. If you rub the comb through your hair and your hair is standing up a little bit, especially on a dry day or in a dry climate, it's clear that's not just friction because now there's something left behind in the hair. Furthermore, the comb has some new property. It's charged. It's statically charged, and if you hold it down next to some little bits of paper or something, they'll dance around a little bit. You can see some curious phenomena.

In order to investigate this, I would like to start at what might seem like an elementary-school level. I really want to go back to the basics, and really the goal here is for you to ask yourself—if I lived a couple hundred years ago and I was trying to make sense of static electricity, what would I do? What kinds of experiments could I do in my kitchen that would teach me something about how the world works? I recognize that most people have learned a lot of the buzzwords and big ideas about electricity and static electricity probably back in grade school, but many times we don't ask—how do we know? Why do we believe these things? I want to lead you a little bit through a path, and it's an over-simplified path but very productive in thinking about how you might go about investigating some new and ultimately extraordinarily rich branch of science—physics.

Let's think about static electricity. One of the Americans most famous for this investigation is Benjamin Franklin. There were a lot of people who were doing this. Ben Franklin was one of a whole crowd of scientists. I don't really think of Benjamin Franklin so much as a scientist. I think of him more in his political role in the early United States, but there's this famous experiment—that we all know about—where he flew a kite during a lightning storm—surely one of the most dangerous and stupid physics experiments in all of history. Benjamin Frankly could certainly have killed himself by flying that kite out in the electrical storm.

He did learn something useful about lightning, and mostly what Ben Franklin was doing was working inside of his laboratory, inside of his house, and he would rub cat fur on amber rods or glass rods with wool. He was just rubbing material objects and noticing that they would move somehow statically charged. He was trying to figure out what the story was.

Here is a simple experiment that you can do, and I can't encourage you enough to go after class and try this out for yourself. Take some Scotch tape, and you're going to waste some tape. Take about a foot of Scotch tape and fold over the ends so that they're taped onto themselves so that you have little tabs so that when you stick it onto something, you can put it off more easily. Take a second piece of tape, make the tabs and lay it down on top of the first piece of tape. Now, you have two pieces of Scotch tape stuck together. Then, take the two of them and lay them down on a flat, smooth table, hopefully a clean table that doesn't have a whole lot of grease or anything on it. There are two pieces of tape, and then do it again. Now you have two duplicate experiments, and it's important that we duplicate the experiment. You will see why.

I'm telling you some steps to follow, but really what I want you to do is play. I want you to think about other experiments that you could do. What if you took a third piece of tape? What if you tried it on a different material? I want you to muck around a little bit so that you can investigate the phenomenon that arises. Otherwise, I'm just telling you the answer, which is really established by people playing little simple games like this. Ben Franklin was fundamentally playing little games like this except he didn't have Scotch tape. Scotch tape is nice because it does tend to exhibit this static electric phenomenon pretty easily. You're doomed if you live in a very, very moist climate. If the humidity in the room is too high, you will find that static electricity experiments don't work so well. You might wait until the wintertime, and it's often drier then. There's a reason why dry air makes static electricity experiments work better, which we'll come to later.

What you want to do is to investigate the phenomenon, and the phenomenon occurs when you rip the pair off the table and then you rip the two pieces of tape apart. It will be immediately obvious that these things are charged. They will act like the clothes coming out of the drier, but they'll stay that way. Just be sure that they don't sort of curl up and stick to your arm or something. You want to make sure they're dangling, and you'll see that they're attracted to you and they're attracted to each other. It's helpful when you're doing this experiment if you take a pen and you just mark the top tape "T" and the bottom tape "B" so that you can remember which tape was on the top and which tape was on the bottom when they were lying on the table. The reason that you want to do that is you will discover that different things happen with the top tape and with the bottom tape.

At first, it may not be obvious. At first, both of them are attracted to you, and they're attracted to each other. It seems as if you've discovered a new attractive force of nature as gravity is an attractive force between masses. Now we have a different kind of attractive force, but the interesting thing, the really surprising thing, about electricity is that if you take two identical pieces of tape—that's why you need the identical set-up—if you take two top pieces and you hold them together, they will bend apart from one another. It's a repulsive force rather than an attractive force. Now you have to scratch your head because something different is going on, and you have to play around a little bit to try to see if you can realize what the systematic underpinning ingredients are in this story. How can we make predictions about who is going to attract and who is going to repel in what circumstances?

What I want to do is think about crafting a model. In the 1700's there was a tremendous amount of intellectual effort. Everybody who was working with static electricity was trying to construct his own model to explain and understand what's going on. It might have been a mental model or a mathematical model. Remember what I mean by models. In old Greek astronomy people were looking at the sky, and they created a model of how the solar system worked. One of the models had Earth at the center, with everybody going around us. A different model had the sun at the center and everything going around it. There was a third model, which Tycho Brahe subscribed to, in which the Earth was at the center, with sun going around us, and everybody else going around the sun. You could imagine very different kinds of models. These are all very mechanical. You could imagine other kinds of ways of thinking about the solar system, and then there is a question. You have your model and I have mine, but let's think of some experiments and look at some data to see whether your model continues to describe the data accurately. If it stops working because we've either come up with a clever experiment or we've just made more careful measurements, then we have to toss your model, and we'll keep refining until we have one that's robust.

This is really the history of electricity and magnetism as well, and in the end the model that I'm going to tell you about is the familiar one that we all learn. What I'm suggesting now is that you could come up with this model, and if you question it or you wonder, why does he say that there are electric charges, for instance, on this piece of tape, which is exactly what I'm going to say. You can ask yourself—let me think of some experiments where maybe an alternative model could be contradicted or verified by the experiment. Maybe you'll come up with alternative model of electricity and magnetism. It would be fairly remarkable since there have now been about two or three hundred years of steady, progressive experimentation and verification of the model that the world is made of objects, we call them atoms, which have electric charge in them. Electric charge is a new word I'm making up, and electric charge is the quantity, the thing, a material thing, which exhibits this new force of static electricity.

Somehow when you rip pieces of tape apart they become electrically charged, and at first you might say, okay, maybe there's one kind of charge in the universe and you can have more of it or you can have less of it. That's a hypothesis. It's a model, and you could test that model and discover that it doesn't work very well because you can't explain why the top and bottom tapes attract one another, two top pieces repel one another, two bottom pieces repel one another, and yet either top or bottom attracts to your shirt. There's a variety of evidence already in this super simple little experiment where I claim that you need to hypothesize two types of electric charge. You have to give them two different names. We could call them the top type and the bottom type if we were just doing the Scotch tape experiment. Benjamin Franklin came up with the names positive and negative. One of the tapes will become positively charged. The other will become negatively charged, but these are just names. He could have called them chocolate and vanilla charge or "X" and "Y" charges. It's nice that he chose the names *positive* and *negative*. We'll talk a little bit about why that's such a productive choice of naming.

Positive and negative already carry the sense of there being two. There are two kinds of charge, and you might ask how Benjamin Franklin knows there's not a third different flavor, a third type of charge. If there were, what would be the experimental ramifications?

Let's try another experiment. We could take a third piece of tape and rip them apart one at a time. We could start messing around, and I claim that there's no intrinsic reason why you couldn't hypothesize a third charge. You just don't bother because you don't need to. There's a principle of physics called Ockham's razor. There was a medieval philosopher, William of Ockham, and William of Ockham suggested that when you're trying to understand the world, and you have multiple possible explanations, go with the simplest one. At least go with it at first, and if it works, it works. Don't go for the complex theory unless you absolutely have to. If you can explain the data of the Scotch tape and you only need to hypothesize *plus* and *minus* charges, great. Let's just hypothesize two different flavors. You'll find you can't explain the attraction and repulsion if you only hypothesize one flavor so we have to move up in complexity. Ockham's razor says keep it as simple as possible. That's probably the right theory. It's an old medieval idea, but it's really become part and parcel of classical physics and, really, all of science. Ockham's razor is often appealed to by scientists, and it really is a philosophical idea rather than a provable scientific idea.

Plus and minus were good names. One reason they're good names is that when you think of plus and minus you think of adding numbers, and you think of adding plus and minus and you have nothing. That's exactly what happens with electrical charges. If you have positive charges and negative charges and bring them close together, they act as though there was no electric charge there. They no longer attract or repel. They no longer exhibit static electricity.

Part of the model that we're building is that these charges are out there in the world. We're not creating them. That's a different model. A different model would be, as you rip things apart, you are making electric charges, but the model that we're going to work with is, they're there. They're already in the atoms, and contemporary language would say the atoms have little negatively charged electrons in them. They're part of the atoms, and you can separate the negative particles from the positive particles. This is a modern interpretation of the old ideas of positive and negative charges, but this model says they're everywhere. The world is filled with positives and negatives. If you rub or rip things apart, you can separate them and notice their effects, but by and large you don't notice them because positives and negatives attract one another. The top tape and the bottom tape attract one another. They're different kinds of charges; like charges repel one another. Two top tapes will repel one another. Two bottom tapes will repel one another. The working hypothesis is that similar charges, the same sign, run away from one another, and opposite charges run toward one another. When they are close together, if you are charged but you're some distance away, then you're both attracted and repelled from the pluses and minuses, and that's why on average you feel nothing when an object becomes neutral.

There's another nice thing about the plus and minus naming convention, which we'll talk more about next time, and that is when I think about forces, I tend to think in the Newtonian way. Newton was arguing that the gravitational force between two masses arises because of the mass. The electrical force is going to arise because of the charge, and Isaac Newton says, how strong is the force? Well, you multiply how much mass you have here by how much mass you have there. We're going to discover the same thing with electrical charges. You're going to multiply how much charge you have here with how much charge you have there and discover that the sign convention also uniquely summarizes this fact of nature that opposite charges attract and like charges repel.

This is a model. It's predictive. You can use this model to guess what's going to happen if you have a third piece of tape. You can explain what happens if you use different materials, and yet, as with gravity, there's this element of frustration involved because I haven't told you what charge is, just as I never told you what gravity is. I just said there is a force between masses, and it's there. It's part of nature. We're going to describe it, and from that description we can then describe an enormous number of phenomena. Anything other than why gravity is there in the first place can be explained. It's going to be the same story with electricity. Classical physics cannot tell you what charge is. It just postulates that charge exists, that like charges repel and opposite charges attract. From that very simple hypothesis and making it quantitative, which we'll do next time, we will be able to predict and understand just about any electrostatic phenomenon and ultimately any electrical phenomenon at all.

For instance, you comb your hair, and it stands up a little bit. What's the deal there? Well, apparently when you rub things, you're separating charges. The comb is becoming one electrical charge, and your hair is becoming the other electrical charge. We're not creating or destroying charges. This is one of the parts of the model. We're just separating them, and that means, depending on the type of comb, if it has more and more negative, then your hair has more and more positive. Now, think about what happens when you have a bunch of hairs, which are very light objects, and they're pinned at one end, free at the other end, and all electrically charged the same charge. Well, our model says like charges repel. All of the ends of the hairs are trying to move away from each other. Now, if you're free to move and there's a whole bunch of you packed together and you want to move away, everybody spreads out. If you have a third dimension, you'll spread up into that third dimension. That is what happens with your hairs. They're just trying to be as far apart from one another as possible, and that's this phenomenon of the hair standing up.

How about at the children's party where you take the balloon and rub it on your shirt—you will discover it's better if you have a wool sweater, and then it will stick to the wall. That seems to be an electrostatic phenomenon. What's the deal there? The wall is not charged. It totally makes sense to me now that when I rub the balloon it would stick back to my shirt, and it does. You can let go, and it's a party trick for the little kids. Everybody loves this, and what happens is the balloon is becoming one charge, quite probably negative, although again it depends on the type of balloon and the type of sweater. The balloon would be becoming negative. Your shirt would be becoming positive, and now you have a highly charged balloon. When you carry it over to the wall and hold it, why would it stick if the wall were neutral? Remember, our working hypothesis is that everything is made of atoms that have electric charges in them, even the wall.

There are little negatives and positives in equal numbers very close to one another in the wall. If you hold this balloon, which is highly negatively charged up near the wall, what's going to happen? Our hypothesis is that any negative charges in the wall that are free to move, will move away, and any positive charges in the wall that are free to move will move toward, they'll be attracted to, the negative balloon. There will be a separation of charges even if they can't move far, and they really can't move very easily in the wall. However, it only takes a small separation because, as we will see, if the charges are further away, the force is weaker. If they are closer, the force is stronger just, as gravity is. It makes total sense that if something is closer to you, the force will be stronger. You can see it with the balloon near you. If it's close to you, it sticks. If it's far from you, it falls to the ground.

That means that the opposite charges are closer and therefore stronger, and so the attraction, which comes from opposite charges, is a little bit stronger than the repulsion that comes from the like charges. The attraction wins, and you have a weak attractive force. It is weak. You'll notice that you have to rub really hard. It sticks really strongly to your shirt, and it just barely sticks to the wall. We're explaining lots and lots of simple physical phenomena by this model. What we'll discover is that anything you can think of that has to do with electricity can be understood in terms of, ultimately, this basic idea.

I've been waffling a little bit about which is negative and which is positive. It's partly a human naming convention. The first experiment was Benjamin Franklin rubbing amber with some cloth. It's clear that there are charges involved, and it's clear that some things attract and some things repel. Benjamin Franklin had to make a choice. Which one am I going to call positive? Which one am I going to call negative? He had no idea at that time about electrons so he had no reason to decide. Today it's a little frustrating in that we think he chose wrong in the sense that the things that move the most easily are what we would have preferred he called the positive charges so it's the positive that runs around in electrical circuits. Alas, his choice was that it's the negative things that ended up being the objects that tend to move more easily in nature. It doesn't really matter. Some things are plus. Some things are minus. If you rub the balloon and you want to know if it's positive or negative, the real way to decide is to walk over to Benjamin Franklin's house, recreate his experiment rubbing his materials, amber and whatever it was, cat fur, that he rubbed them with, look at his convention about which one is negative and then see if your balloon attracts or repels that object. Once you've named one object in the world, all other charged objects will now be perfectly well defined. Of course, ultimately we can come up with rules about if you rub material "A" and material "B", "A" will become negative and "B" will become positive. Again, it doesn't really matter so much. It's the idea of separation of charge that's the more important thing.

I understand this story is very familiar. It's a simple story, and yet what I really want you to think about is how one goes about pursuing and understanding some new phenomenon of nature. In the end, this phenomenon is going to be spectacularly useful, and this is where we build motors from. This is where our electricity comes from. This is where multi-million-dollar experiments originate. It's nice to think about the basic underlying principles and how you would figure it out.

Let me briefly talk about magnets today. We'll come back and talk much more about magnets in upcoming lectures. Once again, I really want to encourage you to play with magnets. It will be a lot of fun. If you haven't played with magnets since you were a kid, you will discover that they're remarkable little things. I prefer that you have better magnets than the kitchen magnets. It turns out that kitchen magnets are complicated. It's hard to do experiments with them. What you want to do is go to a kids' toy store and buy the little plastic rods that have strong magnets at either end. They probably come with some ball bearings to help connect them together. They are awesome, and what you will discover is that magnetic force looks a lot like electric force. You will find that these little rods have apparently two kinds of magnetic—what shall we call it? I don't want to call it magnetic charge because the word "charge" is reserved for electrical phenomenon. We call it magnetic poles just so we have a different distinguishing word, and we can't call them *plus* and *minus* poles because we've used those words. That's too bad because they're good words. We call them north and south poles. Again, you might ask how I know there aren't three magnetic poles. Maybe there are four—north, south, east and west. Do some experiments. Muck around with magnets, and you'll discover that you only need two to explain the entire magnetic phenomenon that you can see.

In some respects, magnets are similar to electricity. In some respects you'll find that they are totally different. One of the ways they are different is that magnets seem to be permanent. Stick this magnet on the fridge, and you can walk away. For the rest of your life you come back and that magnet is still sticking there. If you rub the balloon on your shirt and you stick it to the wall, 10 minutes later it has probably fallen to the ground. Now, there's an explanation for that falling to the ground in terms of our model of electricity. Namely, I've told you that plus and minus charges are everywhere, including the molecules of air and the molecules of water in the air drifting around. If it's moist, water molecules often pick up excess charges, sometimes a little extra negative, sometimes a little extra positive, and as they drift around, if you are highly charged, like a balloon, the little water molecules are pretty good at picking up some of the excess charge, drifting around and carrying it back to your sweater, basically re-equilibrating the universe. This happens, and the more moisture there is, the more easily and quickly it happens. In general, the electrical separation of charge will work itself out. Magnets are somehow different, and we're going to want to investigate in upcoming lectures how it is that magnets are different and what the story is behind them.

Let me conclude this lecture with the statement that you now have a grounding in static electricity, and once we start to put this story together, we'll see that the lightning bolt that Ben Franklin investigated, forces of nature such as friction, pushes and pulls, the push of a solid body, the light from a light bulb, all of the high technology in your house, is all arising from simply the postulation of positive and negative charges and the force, ultimately Newton's law, possibly conservation of momentum and energy, is what we need. That's where we're headed.

Lecture Fourteen

Electricity, Magnetism, and Force Fields

Since Maxwell's time, physical reality has been thought of as represented by continuous fields, and not capable of any mechanical interpretation. This change in the conception of reality is the most profound and the most fruitful that physics has experienced since the time of Newton.

—Albert Einstein

Scope:

In the last lecture, we introduced a "new" force in nature, electricity, and we constructed a model in which there were two kinds of electric charge, positive and negative. *Electric charge* is a term we used to describe the source of the static electrical force. Using our model, we saw that like charges repel each other and opposite charges attract. In this lecture, we'll move away from thinking of electric charge as "action at a distance," as Newton did, and begin thinking about electrical *fields*. This concept will help us to understand electricity better and to reformulate the way we think about gravity. The key player in the origination of the mathematical theory of static electricity was Charles Coulomb, a French scientist and engineer working in the 1780s. Coulomb's experiments with static electricity were similar to those conducted by Henry Cavendish to measure the force of gravity. We'll also look at the work of Michael Faraday, the British physicist who introduced the concept of a force field, and we'll see how this idea allows us to dispense with action at a distance and visualize force as a local phenomenon.

Outline

I. Charles Coulomb (1736–1806) discovered the fundamental mathematical relationship describing static electrical forces in the 1780s.

A. To understand the gist of Coulomb's experiments, imagine charging up a balloon by rubbing it on your shirt. You'll discover that the charges on the balloon tend to stay put. The side of the balloon that you rubbed will be highly charged; the other side of the balloon will not be any more charged than it was to begin with. In physics, the balloon would be called an *insulator*, that is, an object in which charges can't migrate easily. Metals have the opposite property; in metals (*conductors*), charges spread out easily.

B. Coulomb experimented with metal spheres, which allowed him to distribute charges and measure the forces between them.

C. In honor of Coulomb's work, electric charge is measured in units called coulombs. One coulomb (1 C) is a significant amount of electric charge; a balloon rubbed on your shirt might have only 1/1,000,000 C

II. Let's look back again at the force of gravity.

A. Masses cause a gravitational attraction, and the force of gravity is a constant of nature (measured by Cavendish) multiplied by the mass of one object times the mass of a second object, divided by the square of the distance between the two objects.

B. Coulomb discovered that electricity could be described in much the same way, using the idea of charge rather than mass. Multiplying the charge on one object (measured in coulombs) by the charge on a second object will tell us how strong the force is between those two objects. Coulomb also discovered that static electrical attraction or repulsion, like the force of gravity, declines with greater distance between the two charged objects.

C. There are some similarities between the force of electricity and the force of gravity, but the two forces are *not* the same thing. For instance, in considering the force of gravity, we know that all masses attract, but with electricity, both attraction and repulsion take place.

III. Like Newton's law, Coulomb's law was still a description of mysterious "action at a distance," which Newton himself was not comfortable with. The resolution to the problem of two unconnected objects somehow influencing each other was the idea of a *force field*, introduced by a British physicist, Michael Faraday (1791–1867).

A. Faraday was originally a bookbinder. His lack of formal mathematical training compelled him to think of visual ways to describe and understand complex phenomena, such as electricity and magnetism.

B. Let's also use a simple picture to think about the idea of a force field: Imagine a mattress with a heavy bowling ball in the middle; the bowling ball is the *source* of a force.

1. The mattress sags in the middle; that is, the mattress curves down and inward in all directions toward the bowling ball. The curved surface of the mattress offers some potentiality of force. We could verify this potentiality by placing a marble on the edge of the mattress and watching it roll toward the center.
2. Next, we could place the marble at different points all around the mattress and draw arrows to represent the force created by the source (the bowling ball).
3. If we remove the marble, what we have left are arrows drawn on the mattress, which represent a *force field.* Nothing is happening on the mattress, but the possibility is there, and if we placed the marble on the mattress again, something would happen. In other words, a force field exists, whether or not we test it.

C. We can extend this example to a gravitational field.

1. Instead of a bowling ball on a mattress, think of the Sun in the middle of the solar system. If we let go of a test mass near the Earth, it would be pulled toward the Sun. We could again draw arrows in three-dimensional space, which would all point toward the Sun.
2. Even if we remove all the planets from the solar system, the gravitational force field still exists, whether or not a test mass is available to show it.

D. The electrical field is a slightly more abstract way of characterizing electrical forces.

1. A balloon rubbed on your shirt can serve as a source of an electrical field. In this case, we have to use a test charge, rather than a test mass, to see the effects of the electrical field. If both the balloon and the test charge are positive, we would draw an outward-pointing arrow to represent their repulsion.
2. The force felt by the test charge depends on how strong the test charge itself is. This idea is akin to unit prices in grocery stores. A grocery store might have a unit price of $2.00 per pound for beans. This "price field" exists throughout the store, but doesn't tell you how much you

will pay for beans at the checkout counter. The price you pay depends on the size of the bag of beans you choose. In the same way, the strength of the attraction or repulsion felt by the test charge depends on the magnitude of its charge.

E. Faraday simplified the picture of a force field by replacing the arrows with lines drawn away from the charges. For a point charge, this is called a *radial field* because all the lines resemble the radii of a circle.

1. Such a picture gives us an intuitive idea of the physics of a field. For example, the strength of the field is represented in this picture by how closely spaced the lines are.
2. If we had a positive charge at one point and a negative charge at another, the field line diagram would become more complicated.

F. We could also map out a magnetic field. In this case, the tester would have to be another magnet, perhaps a compass needle.

IV. The idea of fields is powerful because it gives us a fresh way of thinking about force that doesn't involve action at a distance. With the idea of fields, we can consider force as a local phenomenon, and we can predict the strength and direction of the force on any object at any point in the field.

Essential Computer Sim:

Go to http://phet.colorado.edu and play with Electric Field Hockey (EFH) and Charges and Fields. EFH will give you a sense of the force on a charge. Keep it simple at first; try to make sense of the connection between the field and the motion. Think back to earlier examples—the force determines the acceleration (not the velocity). Can you arrange charges to make the test charge (the one that can move) do what you want? Notice the Field button at the bottom. When you turn that on, does the result make sense to you? For Charges and Fields, show only the E field (don't bother with V, for voltage, just yet). Add some charges and move them around. What does the "intensity" of the red arrows indicate? (Does that match with what you see when you add an E field sensor? Why is the length of the sensor arrow different from the usual red arrows?)

Essential Reading:

Hewitt, rest of chapter 21.

Hobson, chapter 9.1.

March, middle of chapter 6.

Recommended Reading:

Cropper, chapter 11.

Gonick, chapters 13 and 17.

Questions to Consider:

1. How would you experimentally distinguish an electrical field from a gravitational field? From a magnetic field?
2. You and I both set out to map out an electrical field in the laboratory. I use a test charge of 1 microcoulomb, and you use a test charge of 2 microcoulombs. If we both independently "test" the same point in the room, will we agree *numerically* on the value of the electrical *force* on our test charges? (If not, how will the forces be related?) Will we agree *numerically* on the value of the electrical *field*?
3. An electron and a proton are placed in an identical electrical field. Compare the electric forces on each of them. Compare the resulting acceleration on each of them (direction and "how big," relatively). (Note: Electrons have negative electric charge and are very light. Protons have positive electric charge of the same magnitude as the electron but are very heavy.)
4. What would it mean to say, "An electrical field is real"? What (if anything) does "real" mean, when you're thinking like a physicist? Can you ever see a field? Feel it directly?

Lecture Fourteen—Transcript
Electricity, Magnetism, and Force Fields

We introduced static electricity last time. Electricity refers to a new kind of force that you observe in nature. You rub things together, and some things will attract, some things will repel. Attracting or repelling is a push or a pull. It's a force, just an ordinary force, as Newton is helping us to understand. We can understand the consequences of this force by going back to Isaac Newton, but what we'd like to do is to investigate the force itself. What does it look like? What does it originate from? How does it behave?

We always start with experimental observations, and this is our model. What we've seen from our simple experiments is that there seem to be two different kinds of material objects, two quantities in the world. We call it *electric charge*, and electric charge is just a name we made up to describe how much of this source of this new force there is. The more electric charge you have, the stronger a source of this static electrical force of nature you have. We discovered that there seem to be two kinds of electrical charge. Positives and negatives were the names that we gave them. Those are just human names, which describe to us in a qualitative sense the fact that some charges attract one another, but other charges repel one another. What we've discovered is that if you have the same charge, if you prepare two identical objects that are charged, they'll always repel one another. No matter what, two identically prepared objects will repel one another. Like charges repel one another, and in some circumstances, you can prepare things such that you have both a positive and a negative. In general, when you rub, there's a conservation of electric charge, and one of the objects will become more positive and one will become more negative. Then they will attract one another.

Today we want to quantify this idea, and we want to move a little bit away from this action-at-a-distance idea, this force between two objects, and think about electrical forces in a new way. We're going to talk about electric fields. It's really a kind of force field. We can talk about gravitational force fields or electrical force fields, and in doing so, it's going to help us to understand electricity better. It will help us to reformulate the way we think about gravity. It's moving beyond Isaac Newton in a way, and yet still retaining those classical philosophical ideas of understanding the world in a very simple, pictorial way based on some underlying hypothesis. I'm not explaining what charge is or why charges attract or repel. I'm describing them and describing every electrical phenomenon that you can think of because of that.

The key player in the origination of the mathematical theory of static electricity was Charles Coulomb, a French scientist. He was an engineer, and he did a lot of laboratory experiments with static electricity. He was working and building up from Benjamin Franklin, who was looking qualitatively. Ben Franklin was thinking about the signs of the charges and whether they attract or repel, and Coulomb was trying to measure. How strongly do they attract, and what seems to be the nature of these charges? This was happening in the late 1700's. It was after Ben Franklin, but by the way, it's just before Cavendish and his experiment on gravitational force, and the experiment that Coulomb did to measure the force of electricity is very similar, in some respects, to the experiment that Henry Cavendish did to measure the force of gravity between two masses. Once again, we have large objects. In this case instead of being massive and attracting each other because of gravity, Coulomb's objects become statically charged and attract or repel one another because of static electricity. They're sitting on a similar apparatus, a torsion rod that can twist. The more force you have, the more it twists. It's very sensitive to relatively small forces.

Charging things up can produce relatively large forces, much, much, much larger than the force of gravity between two ordinary-sized objects. Coulomb's experiment is relatively easy compared to Cavendish's, and the idea would be the following: Let's imagine first that you rub a balloon on your shirt and you have it all charged up. What you discover is that the charges on the balloon tend to stay put. If you rub one side of the balloon, that side of the balloon will be highly charged. The back side of the balloon will not be any more charged than it was to begin with. You can test this by holding it close to a piece of charged Scotch tape. Hold the tape near various spots. You can think of lots of simple experiments to verify that the electrical charges tend to be stuck in one place. If you think microscopically, you just have some atoms on the rubber balloon that either have a few extra electrons or a few electrons have been yanked off and are stuck there by chemical bonds. We call the balloon an insulator. An insulator just refers to an object where charges can't migrate easily.

If you take metals, what you'll discover is that charges can migrate relatively easily. If you take that balloon and touch the balloon, maybe rub it a little bit on an ordinary metal sphere, what you'll discover is that you'll rub off some of those excess charges. The metal sphere will become charged, and the charges will spread out as fast as you can tell. It's nearly instant. Remember, like charges repel one another. If you're free to move and you can repel, everybody is repelling everybody. You're going to spread out uniformly. The sphere will tend to become uniformly charged. Now you have to figure out a way of holding that sphere, but if you just have a wooden handle or a little fiber of some kind that's an insulator, then you can hold it and the charges won't run off of the metal sphere. They'll just run around it and spread themselves out.

Mr. Coulomb is playing this game. You can think now of some simple tricks, such as take two identical metal spheres. Charge one of them, and it has some static charge on it. Now, just touch it to another identical metal sphere. Think about what's happening. You have two identical metal objects. The charges that were on one of them are free to roam around. They can now roam over to the other one. They want to spread out so now they're going to evenly spread out over two spheres. Separate them, and you have two metal spheres now, and each has precisely half of the charge that you started with. You can start to make measurements. How much charge do I have? You can divide by two. You can divide by four. You can change the size of the spheres, and you can start to make some measurements now. Take two spheres and measure the force between them. Look and see how the distance between them affects the force, and look and see how much the amount of charge on them affects the force.

Coulomb did these experiments, and in honor of his work, we now measure charge in units. We say if you have a certain amount of charge, we call it one coulomb of electrical charge. One coulomb is a huge amount of electrical charge. If you rub a balloon on your shirt, it will have perhaps one/one-millionth of electrical charge built up on it. One coulomb would be tremendous. If you had a balloon with one coulomb of electric charge on it, it would explode because, of course, all of those charges are trying to run apart and they would pop the balloon.

By hanging these metal spheres next to one another, Coulomb began to investigate the story of the force. Now, you have to realize this is 100 years after the *Principia*, even more than 100 years after the *Principia*. This is deeply influencing Coulomb's thinking. I will remind you of the force of gravity. If you have two masses, they cause a gravitational attraction, and the force of gravity is a constant of nature, which Cavendish ended up measuring multiplied by the mass of the first object times the mass of the second object divided by the square of the distance between them. Coulomb is probably thinking by analogy that maybe electricity is similar. He does his experiments and he discovers, yes, except that it has nothing to do with the mass now. It has to do with the charge. If you have a charge on object one measured in coulombs, a charge on object two measured in coulombs, and you multiply those two charges that will tell you how strong the force is. If you double the charge on one sphere and you quadruple the charge on another sphere, you would have two times four, which is eight times the force between them. That is one of his observations.

Of course, gravity declines by the square of the distance between them, and so Coulomb tested this out and discovered that, surely enough, the electrical, static electrical attraction or repulsion, either way, also goes like the inverse square law. If you're twice as far apart, two squared is four, it's four times weaker force. That's why as you first pull that balloon off of your chest it sticks a little bit, but as soon as it's a small distance away, the force has declined very rapidly because the square of something grows big very fast. Now, you can just carry that balloon away. We're seeing some similarities between the force of electricity and the force of gravity, but, of course, recognize they are definitely not the same thing. You can't say that electricity is somehow also like gravity. For instance, you can have lots of mass but no charge, and you can have an arbitrary amount of charge without noticeably changing the mass. These quantities seem to be very different.

Here is a huge difference—all masses attract. Like masses attract, but like charges repel. There is this other kind of charge so you can have both attraction and repulsion with electricity. Because of that, you can cancel electricity. If you have pluses and minuses close together, an object, even a highly charged object, far away, will not notice anything because it's equally attracted and repelled. You can't hide gravity in this way because there's only positive mass. It just builds up and builds up.

Coulomb has made some discoveries, and he has quantified the rules of static electricity. It turns out that all of the exotic and unusual laboratory phenomena are now explained not just qualitatively but quantitatively. This is clearly the beginning of an electrical science, and everybody is jumping on the bandwagon. Lots of laboratories are trying to study this. People begin to make spectacular strides very quickly. They discover that you can move charges around. Once you realize that charges are free to flow through conductors, and then you can start connecting different objects with conductors and make the electricity flow in the ways that you want. We're at the beginning of a revolution in technology and in science. I will want to talk about that some more. I want to talk about how we design electrical circuits.

Before we do that, I want to think a little bit more carefully about the nature of electric forces. You have to appreciate that Coulomb's law, just like Isaac Newton's law, looks like mysterious action at a distance. You have a charge here and a charge there. They're separated through space, and yet they feel one another. How can they do that? How can a charge somehow reach out through empty space and affect another charge? There's no material connection, and there's still this desire throughout history to be able to think a little bit more mechanically about the universe. It's part of the classical worldview that people are striving toward, to be able to make sense of this action at a distance.

The resolution, the way that people today think about electric forces and gravitational forces, is a delightful revolution in the way we think. It's relatively simple, and it's going to move us away from action at a distance. It's going to allow us to restore the idea of locality. What happens here depends only on what's going on here. I don't really care so much about objects far away. The far away objects are important. They create a condition where I am. Space itself, we will discover, is modified over here, and so if I have an electric charge, it notices the character of the space it's living in and feels forces. We're going to try to make this crazy, abstract idea very concrete. It's the idea of a force field, and once we can think about force fields, it will help us to visualize electric forces and magnetic forces, and it will allow us to move beyond these particular forces.

We can think about any forces in terms of force fields. It is very, very powerful, and even today, moving beyond purely classical ideas, the force field, or the field concept, is extraordinarily fruitful.

The person to whom we pretty much owe this idea is Michael Faraday. Michael Faraday was a British physicist in the 1800's. He wasn't always a physicist. It's an interesting story. He grew up in the English lower class, and he became a bookbinder. He was a very bright man, however, and he read the books that he bound. He was curious about some of what he was reading, some science and physics, and ultimately as a young adult, he went over to a physics laboratory and said, may I work with you? I'm really interested in this.

They took him on, and he began basically as a technician, doing odd jobs. It quickly became apparent that he was smart, and he was catching up and did a lot of self-study. He began to do experiments, and it was his experiments and his theoretical ideas that led us to many of the greatest discoveries in electricity. Michael Faraday was certainly a very influential character. One of the lovely things, though, ultimately about Michael Faraday, was the fact that he didn't have this classical mathematical training, the rigorous, formal mathematics that almost all physicists before and after have had. It forced him to think about these phenomena in terms of pictures. He had to have some way of visualizing, and what a great thing for us to have this simple, visual tool, to simplify or think about, to represent this formidable mathematics ultimately involved in electricity and magnetism.

There are many situations in the history of physics where coming up with a good picture has been enormously valuable. Albert Einstein did this a couple of times in his picture of space-time. A fellow you may not have heard of, Minkowski, came up with a diagram of space and time that allows us to think about the mathematics of relativity in a lovely way. Richard Feynman is famous for his Feynman diagrams, which help us to think about particle physics. Really, Faraday was one of the early leaders, although he wasn't the first. This idea that we've been talking about in Newtonian physics, of drawing an arrow to represent a force, is a simple picture that helps me enormously. When I think of a force, I'm visualizing a little arrow. I can draw it on a piece of paper. A bigger arrow means a bigger force. Having a simple picture can often be enormously productive in terms of thinking about the physics.

Let's think about Faraday's image, and I'm going to take a couple of steps back because it's a slightly abstract story that we're talking about. We'll start concrete. Imagine a mattress. It's absolutely flat. There's nothing going on, and now I'm going to create a source of force on the mattress by taking a big, heavy bowling ball and plunking it down right in the middle of the mattress. Visualize what happens. The mattress sags. It's a very, very heavy bowling ball so the mattress curves down and inward toward that bowling ball. It curves down and inward in all dimensions. There's something going on here even though nothing moves and if I were to look at a spot on the mattress that's not right on top of the bowling ball, it's a curved surface and there's some potentiality or possibility of force. I could verify this by pulling a little marble out of my pocket, because as an experimental physicist I have to have my test mass with me. I put my little test mass down on the mattress and watch it. I let go, and of course, it's going to start to roll downhill toward the big bowling ball. I might put a little mark on the mattress, a little arrow, which indicates which way the force was on the test mass. Now, I go somewhere else on the mattress, and I draw another arrow with a magic marker. I start walking all around the mattress drawing arrows everywhere. If you're close to the bowling ball, the arrows will be big because the force is very big. When you're close, it's much more curved. As you go farther and farther away from that bowling ball, the mattress is less noticeably curved, and the test mass will move more slowly. It will respond less rapidly.

Now, take that marble, put it back in your pocket, and look down. There's no force. There's nothing going on, and yet we have all of these arrows. The arrows represent a force field. That's what I think of when I think of a force field—just a bunch of arrows in space. There's no object there. There's nothing happening there, but something would happen if I put an object down. It would feel a force if there was an object there. That's the basic idea of a force field.

Let's think about a gravitational field now. Instead of the bowling ball on the mattress think of the sun sitting in the middle of our solar system. Now, you could imagine playing the same game. Pull a small object out of your pocket. Put it in some place like near the Earth's orbit. Let go of it, and it will be attracted straight toward the sun so you draw a little arrow. Now you have this three-dimensional space filled with arrows. The arrows are pointing everywhere toward the sun, and they're very big when you're close to the sun and very small when you're far away from the sun. Now, I want you to take away all of the planets from the solar system, all of the objects, all of the masses and just leave the sun. The source of the gravitational field is still there, but there are no other objects. I ask you—is there gravity? It's an ambiguous question. Is there a force of gravity? It's an ambiguous question, but there is a force field. The force field is still there whether you put something down or not. The field is present. Outer space itself contains this force field. Yes, it's an abstract thing. You can't touch it, but you can certainly notice its effect. I'm going to argue more and more strongly that the field is a real thing. It's just a slightly more abstract thing than a particle or an object.

There is one step further in abstraction that we need to make to go from a force field to a general field. First of all, let's think about electric force fields. If instead of the sun I have a balloon that I've rubbed on my shirt and I put it at the center of the room and now I step back so that I have a source of electric field, the balloon is a source of electric field. What do I mean? I mean that I have to pull out of my pocket a little test charge, not a test mass now. A mass won't notice a balloon. A balloon hardly weighs anything, but the test charge will feel the presence of the electrically charged balloon. If my test charge is positive and the balloon is positive, the little test charge will be repelled. I would draw a little outwards arrow. A positively charged balloon would create an electric force field that points everywhere outward.

Now, the remaining level of abstraction that I need to go to has to do with the fact that the force felt by a test charge depends on how big the test charge is or how strong it is. If I put one coulomb down, it will feel a certain force from the balloon, but according to Mr. Coulomb's law, if I put two coulomb's worth of charge down at that same point, it will feel twice the force from the same source. The amount of force depends on the particular test object that I put down, and that's a little frustrating because the whole idea of this field is that we want an arrow that's kind of just a generic arrow that represents the strength of the field. We don't want that to care about what I pull out of my pocket later to test it out.

I have another metaphor for you, which some people find unproductive. Let's think about going to a grocery store, and there is a price field in the grocery store. Suppose I'm interested in beans. This store has a certain price field. It might be $2 a pound, and that $2 a pound is somehow abstractly present throughout the store. I wander around the shelves, and you know how annoying this is—some person picked up the beans and then decided they didn't want them so they put them down on the shelf. It doesn't matter where you find the beans in the store. They're $2 a pound. It's a field that's present throughout the region of space that tells you some information. Does it tell you how much money you're going to pay? Not quite—you go to the front counter and it depends on the size of the bag of beans that you bring up there. If you have a one-pound bag, you'll pay $2. If you have a ten-pound bag, you'll pay $20. It's $2 a pound so the actual price that you pay, like the actual force that an object feels in a force field, will depend on how much it is that you put down there. The field itself is like the unit price. It's just a generic number that tells you information right up front, and it allows you to know exactly what's going to happen at the end when you pay the bill, put the charge down, or put the mass down.

The electric field refers to these abstract arrows, and if you take the size of the arrow and multiply it by how much charge you put down, that's going to tell you the force felt. If you put down one coulomb, that's the standard charge. Then the electric field magnitude will be the same thing in numerical value as the force that you feel. It's like buying one pound of beans. That's how you figure out that the price is $2 a pound.

I'm now visualizing force fields, and you can think of them in a variety of ways. Faraday looked at this picture and said it's a little complicated. I tried to draw it on a piece of paper, and I had arrows everywhere. If I try to draw too many arrows, they start overlapping, and it I shard to look at the picture. He simplified the picture, and this was really the beautiful thing about his representation. You just start at some point, and you begin to draw a line. You start drawing a line in the direction of the local arrow, and then every time you come to a near arrow, you just keep on drawing the line in the direction of the arrows that you hit. Let's think about what this field line diagram would look like for a point-positive charge at the middle, at the origin.

You start at the origin. There's a big arrow pointing outward so you start drawing a line outward and you just keep on going. Every time you hit an arrow it's running straight away, so what you'll end up with are lines running straight away from the center like the spokes of a wheel. We call it a radial field because all of the lines that we draw look like the radius of a circle. Now, how many lines should you draw? Well, that's really up to you. You typically start with an evenly spaced set of points, and you just run them outward. You see where they go, and now you have a nice picture. You can look at the picture and have an intuitive sense of the physics that's going on. For instance, if you look at this point charge, and you think of the spokes of the wheel heading outward from it, the spokes of the wheel are very close together near the charge, and they're spread out as you go farther from the charge. Apparently, the strength of the field is represented in this picture by how closely spaced the lines are. If the lines are close, the field is strong. As the lines spread out, the field is weaker.

Now, what if you were to pick a spot on your diagram where you didn't happen to have a line? That's okay; you just interpolate. If everything looks nice and smooth, if there's line, line, line, and they're sort of evenly spaced, you say, okay, the field is uniform in there and it's the same. You don't really have to draw a line to recognize that there's force field there.

You can do this with electric fields. If you were to have a positive charge at one point and a negative charge at another point, the field line diagram would immediately start to become more complicated. If you think about how you figure out the field lines, just put little test charges down and see which way they want to go. If they're close to the positive charge, they want to run away from it, but if they're close to the negative charge, they want to run toward it. When you think of your picture, you'll see lines leaving the positive charge, arcing around, and heading back in toward the negative charge. You look at that picture, and you can just see where the field is strong, where the field is weak, which way the charge would want to go if you put it there. It's the possibility of force that the electric field is telling you about, and yet there it is on the piece of paper. You can see it throughout all of space and recognize that there is something there in the empty space even before you start laying out your test charges.

I used the words *test charges* because I want to distinguish them from the source charges. The sources are big and they're fixed, and the test charges are so small that they don't perturb the field. As soon as I put down a test charge, if you're being picky with me you'll realize that, well, that's another charge and it has its own field. Fields will add up with the existing field, so that's why we want to keep our test charges as small as possible so they don't really muck up the existing field that we're interested in. We're just testing it. We're feeling it out.

If you have a magnet, you can draw a magnetic field, and you can think about the magnetic field lines. Now, you need some sort of tester, and the tester for magnets would not be an electric charge because electric charges don't respond to a stationary magnet. What you would need is another little magnet. A little compass needle would be perfect because compass needles are really just little magnets. If you have a big magnet and you want to know what the field looks like, take a little compass and just start laying it down and draw an arrow. Which way is the needle pointing? Draw your arrow. Is it jerking really quickly? Draw a big arrow. Again, connect the arrows, draw some field lines, and you'll see the magnetic field pattern.

When you can see the magnetic field pattern like seeing an electric field pattern, it just gives you a sense, even though nothing is happening yet, of what's going to happen. That is really the powerful idea of fields. It's a fresh way of thinking about force. Think about how different this is than Newton's way of thinking about it. If I'm in a room where there is a strong electric charge at the origin and I'm 20 feet away, one way of thinking would be to say, okay, I put down a charge. It's Coulomb's law and Newton's law. There's a formula for the force. It's the charge of the source, times the charge of the test, divided by the distance squared, and it either attracts or repels depending on the relative signs. It's action at a distance. I have two objects, and I need to know about how far apart they are. I need to know all of that. If I'm thinking instead about fields, I can now completely ignore that source. Here I am with my test charge. I put it down. I only need to know what the field is right there where I am. Of course, the field was created by the source so the source still matters, but I no longer pay any attention to it. I can live locally. Once I've drawn my field lines, I know what's going on. I know what the force is going to be, and it doesn't seem like action at a distance anymore. It seems to me that the charge is responding to the field here rather than thinking about it as responding to the source over there.

It's just a shift in the way we think about things. All of the rules are the same. It's still Newtonian classical physics. We're just thinking about forces in terms of how they affect the space around them.

This is a lovely new picture. It's relatively non-mathematical, although Faraday and the hero of our electricity and magnetism story, James Clerk Maxwell, were working together. Maxwell was the mathematician. He put together the rigorous mathematical formulas that explain exactly how these fields were configured, but the picture remains as a nice way, which we'll continue to think about, of telling us about the pushes and pulls. It could be of electric charges, magnetic objects, or for that matter, gravitational objects.

Lecture Fifteen

Electrical Currents and Voltage

...after Faraday was made a fellow of the Royal Society[,] the prime minister of the day asked what good this invention could be, and Faraday answered:

"Why, Prime Minister, someday you can tax it."

—Frequently referenced but probably apocryphal quote

Scope:

In the last lecture, we saw that an electric charge, acting as a source, creates a field in space, to which other charges respond. In this lecture, we'll look at the progress made in the 19th century in applying this understanding, which resulted in batteries, devices for storing charge, and simple circuit elements. Today, our lives are surrounded by electrical (and electronic) devices. The critical distinction (and connections) between voltage and current allows us to make intuitive sense of much contemporary electrical technology and phenomena.

Outline

I. Nineteenth-century progress in the understanding and applications of electricity was rapid and deep. As a direct result, electrical devices have become a significant part of our contemporary lives.

A. The language of simple electrical devices requires understanding the concepts of *circuits*, *current*, and *voltage*.

B. Our goal in this lecture is to build a mental model to answer questions about electricity, such as: What is electricity? How do we guide and control it? How can we understand manifestations of electricity, such as light coming from a light bulb, in terms of our underlying picture?

II. Let's begin with a couple of familiar terms: *insulator* and *conductor*.

A. Electric charges constitute a property of material objects; in other words, material objects have charge on them. As mentioned in the last lecture, in some materials, such as a balloon, that charge tends to stay in one place, and in other materials, such as metals, the charge is allowed to flow. If the charge flows, the material is a *conductor*; if the charge is "stuck," the material is an *insulator*.

B. Air is an example of a material that acts largely as an insulator. If we build up electric charges on a piece of tape, they will tend not to drift through the air. However, if we build up enough static charge in one spot, the electrical force between the charges can eventually push some charges into the air.

1. With enough energy, an electric charge in the air can rip atoms apart. The atoms will later recombine (because the opposite charges attract each other) and release energy.

2. That release of energy can take a number of forms, including a spark of light. This is the phenomenon we see in lightning bolts and in the spark produced when you walk across the carpet and touch another person or a metal doorknob.

C. Another term we need in discussing electricity is *ground*, which we can use in both the standard English sense and a technical physics sense. The ground is a place for charges to spread out and neutralize.

III. How can we harness the flow of electrical charges for practical applications?

A. We need a mechanism to both build up charge and provide a conducting path back to the ground. The result is an electrical circuit.

B. In 1800, Alessandro Volta (1745–1827) discovered a simple setup of materials that served as an electrical "pump"—the first battery. The materials inside batteries separate charges and drive them to one end or the other. The minus sign on a battery means that negatively charged electrons are driven to that end. The plus sign means that positively charged ions are being driven to the opposite end of the battery.

C. To use the battery, you need a connection, such as a wire, between the negative and positive ends; you can then sustain a flow of charges.

IV. Consider another metaphor to think about electrical circuits and the idea of voltage.

A. Imagine a device that lifts up bowling balls, similar to the return mechanism in a bowling alley. Our device lifts bowling balls up in energy. The term *voltage* is roughly a reference to energy in electrical circuits, that is, how high we're lifting the bowling balls.

B. The height to which we lift the bowling balls (to the tabletop versus to the attic) tells us how much work they're going to be able to do when we let them fall back down again.

C. If we pump up an electric charge, we've added energy to that electric charge, just as we've added energy to the bowling balls by lifting them to the attic. *Voltage* is defined as energy per charge, or energy per coulomb. Specifically, 1 volt refers to 1 joule of energy per coulomb.

D. A car battery is 12 volts. This means that 12 joules of energy is required for every coulomb of charge moved from one pole of the battery to the other. Every coulomb that flows through the wire connecting the positive pole to the negative pole yields 12 joules of energy.

V. In addition to voltage, we need to understand the concept of *current*.

A. With our bowling-ball machine, we need to know how many bowling balls we lift and let fall down every second. If we lift and let fall 10 balls per second, we will get a lot more work from the device than if we lift and let fall only 1 ball per second.

B. *Current* refers to the flow rate: how many charges flowing per second.

C. Instead of bowling balls, we can think of a water pump, moving water up into a reservoir. The height of the reservoir (the voltage) is significant. If we can pump the water up to a higher level, we will store more energy per gallon. The other important aspect of this system is how many gallons are flowing per second (the current).

D. The measure used for current is *amps*, named after André Ampère, whom we'll meet in a later lecture. One ampere of current flow is 1 coulomb flowing every second.

E. With real water pumps, the amount of water that will flow each second depends on the pipes that are used. A large pipe allows lots of water to flow through; a skinny pipe introduces more friction—*resistance*—and doesn't allow as much water to flow. The same is true of electricity: A thick wire allows relatively easy flow of electricity.

1. The term *resistor* is used (unfortunately) for materials that are both highly resistive and not so resistive.
2. We can think of resistance as adding friction to a circuit. As current flows through resistors, they dissipate energy and heat up.

VI. When we use electricity, we often need some kind of a pump. We've talked about using a battery as a pump, but the wall socket is also a kind of pump.

A. The two main prongs of a plug are rather like the poles of a battery; they, too, have a plus side and a minus side (which alternate in time).

B. When you plug in an appliance, the electricity flows through the two prongs, forming a complete circuit.

VII. What is dangerous about electricity?

A. High voltage by itself is not intrinsically dangerous. It's like having bowling balls in the attic; if the floor is strong enough (like a battery with a good insulator), preventing the flow of the bowling balls, there's really no danger at all.

B. If a bird lands on a high-voltage power line, it's in no danger. The danger would come if the bird were to stretch its wings and connect the power line with the ground. That opens up a conducting path for the charges to flow through the bird.

C. Multiplying voltage (energy per charge) by current (charge per second) yields energy per second, or power. High voltage, then, is potentially dangerous because it can yield a high rate of energy per second.

VIII. Every electrical device in your house is designed around these basic ideas, and again, we can trace this discussion back to Newton and his concepts of force—pushing and pulling charges—and energy.

Essential Computer Sim:

Go to http://phet.colorado.edu, build circuits with the Circuit Construction Kit, and play with Charges and Fields again (this time, turning on the voltage indicator). Where is the voltage high, and where is it low? The Circuit Construction Kit (CCK) is my favorite sim. You can spend a lot of time with it, getting a sense of, and the connections among, voltage, current, and power. Try to answer question 1 below with the CCK! There are many more sims to play around with, which I leave up to you to investigate, including: Battery Voltage, Resistance in a Wire, Ohm's Law, Battery-Resistor Circuit, and Signal Circuit.

Essential Reading:

Hewitt, rest of chapter 21 and chapter 22.

Recommended Reading:

Gonick, chapters 15–16.

Questions to Consider:

1. Dig up a small bulb (e.g. from a flashlight), a battery, and a single piece of electrical wire. Can you make the bulb light? Once you have done so, try to find at least four different arrangements that light the bulb. How are they similar? What is the requirement for the bulb to light? What can you conclude about how the bulb is "wired up" inside, where you can't see it?
2. It's easy to confuse *voltage* and *current*. What is wrong with a news broadcaster saying, "20,000 volts of electricity flowed through the victim's body"? What language would you use to describe this tragic accident more accurately? In the end, what harms an electrocution victim, the voltage or the current (or something else)?
3. You have two glowing light bulbs. One is rated 100 watts (which means 100 joules/sec), and the other is rated 20 watts. What is the same, and what is different, about these two bulbs and the electrical flow through them? (Is the power the same or different? Current? Voltage?) What can you say about how much "resistance" each one offers? (This is tricky! A clue is that wall sockets in the U.S. are 120 Volts, no matter what you plug into them.)

4. When you get the bill from your power company, what do you pay for, electric power or electrical energy? How are these related? Look at your electric bill. Odds are that it tells you how many kWh, or kilowatt hours, you are paying for (1 kilowatt means 1000 watts, which is 1000 joules each second). Note that kWh is *not* kilowatts *per* hour! It is kilowatts *times* hours. If you can make sense of why your power company charges you for kWh, you'll have mastered the big ideas of power and energy from this lecture!

Lecture Fifteen—Transcript
Electrical Currents and Voltage

So far, we've been thinking about static electricity and a little bit about static magnetism. We have been talking about the field. The electric field is a new and very abstract concept. If you take a charge, some object that applies electrical forces on other objects that are charged, what we think about is that the source, or charge, is creating an electric field around it. Space is filled with an electric field, and other charges elsewhere respond locally to that field. We've shifted our idea of thinking about force from action at a distance to creating a field and then responding to a field. It's a nice idea. It's an abstract and somewhat formal idea. It's happening in the 19th century. In the 1800's, people are rapidly developing this formalism. They're beginning to understand the ideas of electricity and magnetism, and something very important happened—a big change in the way physics was evolving. Applied physics began to appear. It was obvious to just about everybody working in the field that this electricity stuff was going to be useful. There are going to be all sorts of things that you can do with this electricity thing. It starts off as a curiosity and then a laboratory experiment, and then a laboratory idea, but very quickly, it evolved into something practical, something that, of course, is now part of our lives in almost every aspect of our lives. The flow of electricity is critical. Today I want to think a little bit about the applications of these abstract ideas of electricity and think about these practical applications, such as the toaster oven or the light bulb. You could imagine going further and talking about microwave ovens or computers, and that's just adding some levels of complexity. Today I want to focus on the simpler electrical circuits and try to think about them and make sense of them. I'd like to have a mental model of what's going on in these ordinary, every day electrical applications that connect to the laws of physics, to Newtonian ideas, forces and energy, and now to fields.

We're going to introduce a few new words. We're going to talk about *electric circuits*. We're going to talk about *electrical current*. We're going to talk about *electrical voltage*. These are new ideas, in a sense, although you'll find that all of them refer ultimately to the ideas that we've been talking about. Electric voltage, for example, is really just a way of talking about energy questions when you're dealing with electricity.

At some level, at some underpinning level, Newton's laws are telling this whole story even though Newton never thought much about electricity or magnetism. He didn't develop this theory. Nonetheless, we're still really doing Newtonian physics. It's just the consequences of Newtonian physics. The question that you might ask—thinking like a Newtonian scientist—is, what is electricity? How do you guide it? How do you control it? What does it mean when you observe certain physical manifestations of electricity, like a glowing light bulb, in terms of our simple, underlying picture?

Now, after one lecture can you go and fix an electrical appliance? It depends on how much background you have. You might have some sense about what's dangerous and what's not. I guess I would warn you that a little physics knowledge can be a dangerous thing. Think about safety issues if you go and try to fix the toaster oven, for example, unplug it first.

I would at least like to talk about the underlying story. Let's begin with a couple of words that we've already casually introduced, which are insulating or insulator, and conducting or conductor. The idea here is that electric charges are a property of material objects. Material objects have charge on them. A piece of tape ripped apart from another piece of tape has some electric charge on it, and what we learn is that some materials make that charge stick in one place, and other materials, such as metals, allow it to flow. If the charge can flow, then we have a conductor. If the charge is stuck, we call it an insulator. These are decent words. I think they lead you to the correct intuitions about what's going on. You could ask, what determines this? Let's think about air, for example. Air is a material substance. It's composed of molecules, largely nitrogen, and it's not very dense. It's certainly not a metal, and it's certainly not a solid. What is it? Is it an electrical conductor or an electrical insulator? It's a slightly tricky story.

By and large, air is a decent insulator. If you start building up electric charges on a piece of tape, for example, they will tend not to drift through the air. It takes quite a lot of time, and it's a very, very slow process for the humidity in the air to take those charges away, and if the air is very dry, the air is a pretty darn good insulator. Charges won't spontaneously flow through the air, but they can. If you build up enough static charge in one spot, the coulomb repulsion, the electrical force between the two charges will push them harder and harder, and at some point one of them is just going to be pushed hard enough that it will start flying through the air. What happens next?

Well, if it doesn't have a whole lot of initial energy, it will probably bump into an air molecule and come to a halt, and it still won't be able to travel very easily. If, however, it has enough energy, if you have enough static charge built up so that you can give it a good shove, then your little electric charge can start flying through the air. It's going to knock into the air molecules as it goes, and it's going to impart some energy to those air molecules. What happens if you have an air molecule and you throw some energy into it? Well, it's composed of charged particles. You are attracting some of those charged particles. If it's an electron that's flying by, the positive part of the atom is attracted. The negative part of the atom is repelled, so you're ripping apart the atom. The energy to rip it apart is coming out of the original electron, and now this atom is ripped apart and there are some positives and negatives floating around. They want to come back together so they fall back together, and they will release energy. It took energy to pull them apart, and the conservation of energy says they will release energy as they fall back together again. That release of energy can come in various forms, one form of which might be light.

This is the phenomenon of sparks. Lightning bolts are nothing more than a very huge charge being developed up in the cloud, and then at some point it builds up so much that you have a path for the electric charges to flow all the way down to planet Earth. Along the way, they're whacking into the air molecules, splitting them apart, they recombine, they glow, and you see this bolt, which is really the path of the electric charges. You've made the air into a conductor in this aggressive and violent way. The charges flowed to the ground, and the word "ground" means planet Earth. In the world of electricity, *ground* means any place where charges want to go to. Remember, electric charges are repelling one another. They want to spread out as much as possible, and so the word *ground*, in the electrical sense, refers to a large place where charges can spread out. If I, man, anthropomorphize, and I do this all of the time with electric charges, they want part from one another. They want to go to the ground so that they can spread out and move away from each other.

When you're walking on the carpet in the wintertime and the air is very dry, you pick up static electric charge from the carpet. It's on your body. It can flow a little bit. Your body is not a perfect insulator. In particular, if you're sweating a little bit, salt water tends to conduct electricity better than dry skin, and so you'll build up some charge on your skin and at some point it may discharge. It wants to go as far apart as possible. It would prefer to go to the ground, and one way to go to the ground would be through something that conducts electricity easily such as a metal object, a doorknob. That's why you tend to have the shock, the little spark. You can see the visible trail of the electrons traveling through the air as you are discharging from your body to this ground. The word *ground* doesn't necessarily have to mean planet Earth. The doorknob can serve as the ground. As far as electrons are concerned, it's a big space for them to spread out.

What are we going to do with this static electricity? It doesn't really help us all that much in a practical sense if we can just build up charges. What we really what to do is to create a device where charges will flow, because if they're flowing, then the electrical force that they apply can push on something and do something for us. If you want to apply this physics, you have to think about a mechanism of some kind, which will create a flow of electric charge. You could think of a crude and simple mechanical way of doing this. You could rub the balloon on your shirt, charge it up, move it over somewhere, touch it onto a metal object, and the metal object will be all charged up. If you do this for a while, the metal object will become more and more charged up each time you touch it, but of course, the more charged it is, the harder it is to push that balloon toward it because like charges repel. At some point, you would just give up. The metal object is charged as much as it can be, so you would stop. We have this pumping mechanism, and if there's no mechanism for the charges to return, then we are stopped.

What you need now is to connect that metal sphere or object, whatever it was you were charging, back down to the ground or in this case, the source of electrical charges. You can think of ground as the sink for electrical charges, but you can also think of it as a source for electrical charges. If they're free to run away, they are also just free to run. They can run toward you just as easily.

What you would need, if you wanted to create the circuit, is a pump. That's a metaphor. You need a mechanism to separate charges and move them to a place, and then you need a return, conducting path. Usually a metal wire will do it. Now we have an electric circuit. If you have the pump going—it has some little Rube Goldberg device where I rub the balloon, I touch the metal sphere, the metal sphere is connected to me with a wire—the charges will go through this circular path. The balloons are the pump, and then the wire completes the circuit.

If I keep the pump going, then we can keep this electric flow going, and you can imagine that you might be able to use this for some purpose because you have electricity flowing through a wire. If you want to make this practical, you're not going to be able to do it with rubbing balloons. You need some device that does this for you, some electrical pump, and the person who discovered this was Alessandro Volta in 1800 in Italy. Alessandro Volta was mucking around with different materials and discovered some remarkable chemistry. It is just fortunate that nature does this for you. There are certain materials that you put together, different kinds of metals and materials that you stack up, and internally some chemistry happens. The chemistry is a microscopic motion of molecules and charged objects. It physically moves electrons over preferentially to one side of this material object. We're not going to worry about the microscopic details of this battery, which is what you call this kind of electrical pump. This battery is just driving electrons to one side. If you buy a little D cell now, it just has some materials inside of it that separate charges. One side will be labeled plus. One side will be labeled minus, and what the plus and minus mean is that the negatively charged electrons are being pumped toward that minus side. The positive ions, the positively charged particles are being driven toward the positive side.

If the battery is just sitting there on the shelf, it's like the original story with the balloons. You do separate some charges. There's a little bit of excess negative on the negative side and a little excess positive on the positive side, but at a certain point it stops pumping. It's no longer possible to keep on adding more and more negatives. The thing isn't going to spontaneously explode, and what you want in order to use that battery is to take a wire and connect the wire from the negative to the positive. It doesn't really matter which way you think about it. Charges will flow, and now you can keep this going forever. The battery inside drives the charges to the negative side. The negative charges run away from each other. They go back to the positive side, and now you have this lovely circuit.

Let me think of a metaphor, because I want to think of this in a variety of ways. I want to think about pushes and pulls like Newton's forces, but I really want to think about energy and the flow of energy. Let me think about a mechanism, and this mechanism is going to be mechanical. I'm going to have a pump that lifts, for example, water; or just to exaggerate a little bit, bowling balls. The bowling balls are going to be lifted up as at the bowling alley, where there's a belt that lifts them up in the air up to a higher point. It lifts them up in energy. In space, lifting some massive object up a distance is also lifting it up in gravitational potential energy. We're going to want to think about this in electric circuits, and the word that we're going to use in the electrical circuits is *voltage*. Voltage is a reference to energy. It's how high you lift the bowling ball, but voltage is a little bit subtle. We're going to have to come back and really be careful about our definition of the word. It's not a synonym of energy. It's just related to energy. Take these bowling balls and lift them. How high you lift them tells you something important. It tells you how much work they're going to be able to do when you let them fall back down again. If you lift the bowling balls up into the attic, they can do more work for you when you let them flow back down again than if you only lifted them up to tabletop height.

In order to do something useful, you need the pump and you need a channel. You need a mechanism for them to flow back down again, just as in the electric circuit you need the battery and you also need the conducting wire to bring them back down to where they started. Let me define voltage for you now. If you take an electric charge and pump it up, you have added some energy to that electric charge, and I'm going to define voltage as the energy per charge.

Think about that for a second. It's not exactly the same as energy. It's the energy per coulomb. If I lift one coulomb up, I will have put a certain amount of work into it. I've given it a certain amount of energy. If I lift up another coulomb, I've done the same amount of work again. Overall, it took me twice as much work. If I have twice as much charge up in the attic, I've done twice as much work, but the height of the attic is a fixed, constant number. The voltage is the energy per unit charge. It's like the height of the attic. No matter how many bowling balls you put up in the attic, the height is a fixed number. Once again, it's like a unit price. No matter how many pounds of beans you buy, the unit price is still $2 a pound, whatever that number might be in the store you're at. As you lift more and more electric charges up to the end of the battery, they all have the same unit energy, energy per charge, and that's the voltage. One volt, named in honor of Alessandro Volta, refers to one joule's worth of energy for every coulomb that you put up there. If you lift one coulomb of charge up and you put one joule of energy into it, you have a one-volt battery. Now, if you put another coulomb up there, it's still a one-volt battery, but now you have two joules of energy. The more charges you put up there, the more energy you have even though the voltage, the unit price, is still staying fixed.

If you have a car battery, it's a twelve-volt battery. What does that mean? It means that every time you move a coulomb of charge from one side, one pole, of the battery to the other side of the battery, which happens by chemistry inside, every coulomb that you move took twelve joules of energy. You'll have it back when you connect some wire from the positive to the negative on the car battery. Every coulomb that flows through that wire is going to give up twelve joules of energy. If you have two coulombs that go through, you'll have twenty-four joules. If you have ten coulombs that go through, you'll have 120 joules. The more charge that flows, the more energy you'll have out of that battery, but it's always still just a twelve-volt battery.

Voltage is very important because energy is very important. It tells you how much work these charges are going to do, but it's not the entire story. If you think back to the bowling ball metaphor and you want to know how much work you can do, you certainly need to know how high up you're storing those bowling balls. The higher you store them, the more work each bowling ball will do when it falls back down again. On the other hand, it also matters how many bowling balls you lift and then let fall down every second. If you lift ten bowling balls per second, you're going to have a lot more work done than if you only lift one per second, even though they're all going to the same height.

What we're now talking about is how many per second, and that we call the electrical current. The word *current* refers to the flow rate, how many charges flow every second. I like the name current. It makes me think of water, so instead of bowling balls, I can think of a water pump. It's pumping water up into a reservoir. The height of the reservoir certainly matters. If you can pump it up higher, you're going to store more energy per gallon. That's one of the things that definitely matters, but the other thing that matters if you're trying to run a water mill, is how many gallons per second are flowing. The two things are separate and independent. You can have a little pump that pumps one gallon per hour up to the particular height. You can have a big pump that pumps a thousand gallons an hour up to that same height, and those will be different situations even though the height is the same. Or, vice versa, you can have one gallon per hour going up ten times as high, and again the situation will be different. We need to keep track of both current and voltage.

If you want to measure volts, one volt is one joule per coulomb. If you want to measure amps, the unit of measure we use for current, which is named after a fellow named André Ampère, who I haven't talked about yet. We'll come back to Mr. Ampère, who did a lot of work studying electric current flow and in particular, its connection with magnetism. We haven't quite come there yet. One ampere of current flow is defined as one coulomb of charge flowing past you every second. Think about that for a second. One coulomb of electric charge is a lot. If you put a coulomb of charge on a balloon, it would explode. One coulomb per second, which is one amp of current flow or one ampere, is a huge current, and yet we use amps in our houses every day. There's a lot of current flowing through our houses.

When I'm thinking about this, sometimes, and in fact, almost always, when I think about electricity, I have this very mechanical metaphor in my mind of pumping water up to a certain height. That's like the voltage, and then I ask how many gallons are flowing each second? That's the current. Now, in real life, with water pumps, how much water will flow each second will depend on the pipes that you use. If you have a big, fat pipe, you can have lots of water flowing back down. If you have a little, skinny pipe, there will be lots of friction, there will be lots of resistance, and then you won't have so much flow. The same thing happens with electricity. If you have a big, thick wire, you can have a relatively easy flow of electricity. When you look at the wires coming into your house where all of the electricity has to come in at one point, you'll probably use large gauge, big fat metal wires, but going out to an individual light bulb, you can use skinnier wires and save some money. It has to do, in part, with how much current you need to flow every second for the particular application.

When we have a metal conductor—and that's a good name, because I think of a metal as allowing electricity to flow; alas, the name that we typically give to pieces of metal is not conductors—we call them resistors. Resistor is a generic name that can refer to a beautiful piece of metal that is hardly resisting at all so it has a very small resistance, or it could be some material that's been contaminated with all sorts of other atomic ingredients, molecular ingredients that are resisting the flow of electricity. You can have a resistor that's very resistive or a resistor that's not so resistive. It's like having a big, fat pipe or a little, skinny pipe. You can choose. You can design materials to have some amount of resistance, and in many applications, it's precisely what you want. You might think, why don't you just go and find things that are the best conductors possible? You could in principle. There are materials called super conductors that have zero electrical resistance. Current just flows infinitely easily through them, but we don't use those in our everyday lives. We always use real resistors.

One of the nice things about a resistor is that it does have friction. It's just friction for electrical charges instead of mechanical bowling balls or flowing water, and resistance always ends up meaning heat. Many of the applications of electricity in our ordinary life are simply heat. You have an electric heater. You have a toaster oven. Even the light bulb is really glowing because it's so darned hot. It's glowing first red hot and then yellow and white hot. That's the problem with regular incandescent bulbs—most of the energy is leaving in the form of heat and not in the form of light. It is certainly an enormously practical application of this idea, having some sort of battery and some sort of circuit so that the water or the electricity can go up the pump and down the conducting path. It gives up some of its energy. The resistor allows it to give up its energy. If you have a 12-volt car battery, in principle what you can do is you can gain 12 joules out of every coulomb that flows by and use those joules any way you like. It's energy, so whatever energy can do for you. It can turn a crank. We haven't talked about how we can use electricity to move an object physically. That's going to be the topic of a future lecture. Obviously, that's the most exciting and practical application you can think of for electricity, but also heating things and lighting things. It is a very, very useful idea.

In general, when you're working with electricity you always need a pump, and the battery is a great thing but we don't always use batteries. Sometimes you plug something into the wall. If you look at the wall socket, there is the old-fashioned kind that only has two plugs, and that's the kind I want to think about. The more modern plugs have a third socket, called the ground, which is really a safety device. It's not really an integral part of the circuit. You can think about the two main prongs, and it's just like the two poles of a battery. The battery has a plus and minus. The electric wall socket has a plus and a minus side. The big difference between the wall plug and the battery is that somebody far away is reversing the battery behind the wall plug 60 times a second. One side is plus and the other side is minus, and then they switch them, minus/plus, plus/minus, minus/plus. The electric current first flows one way. Then it flows the other way. Then it flows one way. Then it flows the other way. It's a subtle difference, and we don't really have to worry about it so much. With all of the physics, microscopically it really boils down to the same thing. When I think about the wall plug, I pretend in my mind that it's a battery, and for many, many purposes that's perfectly fine.

If you plug something into an appliance, take a look. There are two prongs because, of course, you have to have a flow. The electricity has to come in one side, do something, go through the resistor and go back out the other side. If you don't have a circuit, then think about the water pump analogy. If you don't have a path for the water to flow back down again, it just builds up and finally the pump shuts down and you have this tank filled with water or you have an attic filled with bowling balls and they just sit there waiting. You have to have the closed circuit in order to have something interesting happening.

If you're thinking about electrical safety, you ask, what is dangerous about electricity? Well, there is energy available. If you stick your fingers across the plug in the wall socket, that's a bad thing to do because it's 120 volts in the United States. It's like having 10 car batteries all stacked together and touching both ends simultaneously. If you're at all sweaty, then there will be a nice conducting path through your fingers, and of course if the electricity flows through you, you're a big resistor because your body is not made of metal and you're not a great conductor. All of the energy of those charges is going to be given up in your body. It's going to heat you up, and that's a horrible feeling. That shock that you feel is the energy being deposited, and the less resistance that you have, the more charges will flow every second. You can sustain a really nasty shock if you're, for instance, immersed in water because the water allows more electricity to flow, more current flowing.

I want to think about this in a little bit more detail. Think about damage or danger of electricity. Some people think it's just the voltage. High voltage means high danger. Well, not exactly—sort of. If you have an attic full of bowling balls, it's maybe dangerous but not if you have a good, strong floor. If there's a good insulator preventing the flow of those bowling balls, then there's no danger whatsoever. You can have as many of them up there as you want. If a bird lands on a high voltage power line, it's in no danger because the bird lands on the high voltage power line and it's like the bird landing up in the attic. It's just walking around amongst all of those bowling balls. It's perfectly happy. The danger would be if the bird were to stretch its wings and connect the high voltage power line with the ground. Now, you have a conduction path through the bird, and charges will flow. It's as if you opened a pathway for all of the bowling balls to fall out of the attic, and now you can really cause some damage.

How much damage? Well, we can be very quantitative about this. Voltage tells you energy per charge. That's not all you need to know because you also need to know how many charges. If you have a certain energy per charge, that's the voltage, and you multiply it by how many charges flow every second; remember that's what current was—if you multiple voltage times current, that's energy per charge times charge per second, you have energy per second. Remember what energy per second is; it's power. The energy flowing through something is the product of voltage times current. They both matter. High voltage is definitely dangerous because multiplying a big number times just about any other number gives you a big answer unless you multiply by zero. High voltage is not dangerous if there's no mechanism for current to flow. As soon as you create the path, sticking your fingers in the plug, now all of the sudden you're going to have both the voltage, it's always 120 volts with the wall plug, multiplied by however much current you can draw, and that's going to depend on how resistive you are. That's going to tell you the power.

Every electrical device in your house is designed around these basic, simple ideas. There's voltage supplied. That's energy per charge. Some current will flow. You always need a loop. You will always have a circuit, and you can look at any device such as your toaster oven and can see the wire coming in. You can see the wire going out. People are often very clever about putting those two wires close together and packaging them so it looks like a single wire coming from the wall. Look carefully. If the toaster is totally broken, rip it open, and you will see there are always two wires because you have to have that circuit. What we're talking about now, voltage and current, sounds like a whole new story. You might think it's totally disconnected, but in fact, really we're just talking about pushing and pulling charges. It's Mr. Coulomb's law in action. We're thinking about energy. That's all voltage is. It's just a fancy word for energy per unit charge, so we always, when we think about these novel things, in the 1800's, the 1900's, and even today, ultimately go back to Isaac Newton, his concepts, and the principles of energy and energy conservation as well. This is the way I tend to think about electricity.

Lecture Sixteen

The Origin of Electric and Magnet Fields

Aye, I suppose I could stay up that late.
—James Clerk Maxwell, on being told on his arrival at Cambridge University that there would be a compulsory 6:00 a.m. church service

Scope:

Despite all our wonderful technologies, electricity and magnetism are forces we only rarely experience *directly* (e.g., static cling or kitchen magnets). These two forces are distinct but intimately connected. We can create (electro)magnets out of completely nonmagnetic materials, making use of pure electric currents. And we can produce electrical currents by spinning magnets near wires. In this lecture, we zoom in on the *sources* of electric and magnetic fields and their myriad connections, leading to a deeper understanding of the unity of electromagnetic physics.

Outline

I. Let's begin this lecture by talking about magnets; specifically, we'll look at the commonalities and differences between electricity and magnetism.

A. Magnets have two poles, north and south. As with electric charges, like magnetic poles repel each other and opposite magnetic poles attract.

B. This phenomenon looks like action at a distance, but if you hold two magnets close together, you can almost feel the force field operating between them. A compass needle, which is a small magnet, responds to the magnetic field, just as a test charge responds to electrical fields.

C. You can map out a magnetic field by placing a piece of paper over a magnet and sprinkling iron filings on the paper. The filings will arrange themselves in a field line pattern, showing magnetic fields looping from north to south poles.

II. What are the differences between electricity and magnetism?

- **A.** You can charge materials up electrically by rubbing them, but the same thing doesn't work for magnets. Most materials are not magnetic.
- **B.** Magnets stay magnetic for a long time.
- **C.** Compass needles do not deflect in the presence of electric charge. In other words, magnets are neither attracted to nor repelled by electric charge. This alone tells us that electricity and magnetism are two distinct forces of nature.
- **D.** By the same token, electric charges are not attracted to magnets. The Earth is a giant magnet, which causes all compass needles to point north. But there is no attraction of electric charges to magnets.
- **E.** Electric charges can be separated, but magnetic poles can not.
  1. I could charge up a balloon, hand it to you, and you could walk away with it; you would then have an isolated electric charge.
  2. If you cut a magnet in half, however, you would not end up with an isolated north pole or an isolated south pole; you would have two smaller magnets, *each* with a north pole and a south pole.
  3. If you *could* isolate a magnetic pole, scientists would call that a *magnetic monopole*.
- **F.** The bottom line is that electric charges interact with other electric charges, magnets interact with other magnets, and masses interact with other masses. Thus, all three forces at work here—electrical, magnetic, and gravitational—seem at first to be independent and unrelated.

III. In the early 1800s, Hans Oersted (1777–1851) discovered, while preparing for a classroom lecture, that electrical currents can produce magnetic fields.

- **A.** This discovery was a surprise. Electric and magnetic fields were assumed to be separate and distinct; Oersted found that flowing electric charges in a simple circuit create a magnetic field. Oersted's simple, reproducible experiment generated significant interest.
- **B.** Merely closing a switch to allow current to flow creates a magnetic field, and can lead to practical applications.

IV. The French mathematician André Ampère (1775–1836) began to formulate a law to explain magnetic fields.

A. As we said earlier, static electric charges (that is, charges that are not moving) create electrical fields. The field lines in our earlier picture emanate in straight lines from (or toward) electric charges in the center.

B. With magnetism, you might think that the pattern of field lines created by an electric current would be similar; that is, you might expect to see radial magnetic field lines pointing away from a current-carrying wire. However, magnetic field lines run in circles around the electric current.

C. In thinking about this phenomenon, keep in mind that electric charge is flowing, but the field is static. Ampère worked out the mathematics to predict the magnetic field generated from any current in any strength.

D. You could run an experiment at home to create a magnetic field, although you would have to exercise some caution. You would also have to take into account the fact that the Earth has its own magnetic field; you would be superposing your magnetic field on the one that already exists on the Earth. What would be the result of this superposition?

1. Recall Galileo's superposition principle applied to forces: Two forces acting in opposition will cancel each other out; two forces acting in the same direction will add up.

2. If the magnetic field you create is aligned with the Earth's magnetic field, the resulting magnetic field will be stronger. If the magnetic field you create points in the opposite direction of the Earth's magnetic field (and is equally strong), the result is no magnetic field at all.

V. One very strange aspect of electricity and magnetism is that static electric charges don't generate magnetic fields, but moving electric charges do.

A. If I charge up a balloon and place it in a room, no magnetic field is generated. But now you enter the room in a slightly different reference frame, moving steadily through the room on a cart. From *your* reference frame, you are at rest, and the room is sliding by you at 2 meters per second. To you, the balloon is moving, which means that electric current is flowing, and a magnetic field is created.

B. Electric and magnetic fields are very real and intimately connected, but the value and nature of the fields are dependent on the observer. Ultimately, physicists will give the name *electromagnetism* to this one force of nature, which seems to have two sides to it.

VI. Let's return now to applications.

A. A certain amount of current will create a certain magnetic field strength. Doubling the current will also double the field strength. You can make a very powerful magnet simply by running current through coils of wire.

B. When you insert a key in your car, you complete a circuit that starts electricity flowing through a coil of wire (the solenoid). The resulting magnetic field is strong enough to pull an iron rod, engaging the starter.

VII. Where does the magnetic field of an ordinary kitchen magnet arise from? No obvious current is flowing to generate a magnetic field.

A. In fact, there is current in this situation, but it's microscopic. The full story requires modern quantum physics, but we can make basic sense of it with a simplified classical physics picture.

B. An atom of any material has a positive, heavy nucleus and negative electrons in orbit around it. The moving electric charge of the electrons constitutes a current. This current creates a tiny magnetic field; thus, atoms themselves are tiny magnets.

C. If nearby atoms are randomly oriented, the magnetic fields they produce will cancel. That's why ordinary materials aren't magnetic. But in special materials—such as iron crystals—the magnetic fields of the atoms align; adding these microscopic fields up creates a macroscopic magnetic field.

VIII. Faraday discovered moving magnets generate electrical fields, perhaps the most practically important discovery in the history of physics. Rotating magnets at a power plant are the source of the electrical field that pushes electrons through your toaster oven or computer.

Essential Computer Sim:

Go to http://phet.colorado.edu and play with Faraday's Electromagnetic Lab. There are tabs for different experiments. Look at the magnetic field; do you understand this representation of the field? The Pickup Coil variation lets you directly study Faraday's law of induced currents. In Electromagnet mode, turn up the voltage and see if you can visualize the circles of magnetic field around the coils.

Essential Reading:

Hewitt, chapters 23–24.

Recommended Reading:

Cropper, chapter 12.

Gonick, chapters 18–19 and 21–22.

Questions to Consider:

1. Opposite magnetic poles attract (north attracts south). A compass needle is a tiny magnet, and (by convention) we label the pole that points toward geographic north (Canada) the "north" end. (Seems reasonable; magnetic north points you geographically north!) Now think carefully (draw a picture) and decide which magnetic pole of the giant magnet that is planet Earth is the one located in northern Canada. (The answer may surprise you.)
2. Can a constant (steady, unchanging) magnetic field set into motion an electron initially at rest? Try to explain your reasoning carefully.
3. Particle physicists send high-energy microscopic particles through *bubble chambers*, where they leave a trail of bubbles as they pass, enabling physicists to track their motion. There is always a strong magnetic field in the bubble chamber, and some particle tracks form spirals, while others are straight lines. What can you conclude is different about these particles?
4. Iron is a magnetic material, but it is not always a "magnet." Most pieces of iron have no poles, but a magnet sticks to it. (Your refrigerator has iron in it that is not itself magnetized, but a

magnet will attract to it.) Let's say I hand you two heavy, identical chunks of iron, one of which is magnetized, and the other is not. Think of at least three different ways you might figure out which one is the natural magnet and which is unmagnetized iron.

Lecture Sixteen—Transcript
The Origin of Electric and Magnet Fields

Despite all of the wonderful technologies that we live with that use electricity and magnetism in our everyday lives, we tend not to have a lot of personal or direct experience with these forces of nature. We have a lot of experience with gravity. We have a lot of deep intuitions about how the gravitational force works, but electricity and magnetism are inevitably more abstract so in order to make sense of them we tell stories and we come up with metaphors, analogies and representations. Today I want to continue to think about electricity and magnetism, these two forces of nature, and try to make some more sense about what they have in common with each other and with gravity, and what is different about them. In particular, I want to start to lead us toward the question of where did they come from? Once we begin to see this, we will see that electricity and magnetism are connected to one another. It's going to lead us down the path where we're headed for the next couple of lectures, in which we understand the unification of electricity and magnetism and many more pieces of physics together.

I'd like to begin today by talking a little bit more about magnets. We introduced them a couple of lectures ago, and I hope you've looked for some magnets to play with, just to muck around with. Magnets can be found in nature. The ancient Greeks found loadstones, which were magnetic. They would attract one another and in a different orientation repel one another, just like the toy magnets you can buy at a store today. The Chinese discovered magnets many thousands of years ago. The Chinese had compasses before the western world had developed this technology, and you'll discover when you start looking at and playing with magnets, that they have some unique properties. First of all, you have to find one. Once you've found one, then you can create a new one. The easiest way to do it is with iron. Go to the hardware store and buy an iron nail. You can rub it with a magnet, and as you rub it, you will find that the iron nail becomes magnetized. It becomes a magnet.

Let's think a little bit about magnets and how they work, and I'd like to do the compare-and-contrast thing with magnets and static electricity. We already did this a little bit. I'd like to think a little bit more deeply about their differences, so that when we see the commonalities, we'll recognize how remarkable it is that electricity and magnetism have anything to do with one another whatsoever.

If you look at a magnet, you will find in general that it has two poles. Now, if you look at a kitchen magnet, you may be hard pressed to figure out where the North Pole is and where the South Pole is. That's why I encouraged you to buy the rod magnets. If you look carefully, you will discover that there are north and south poles on the kitchen magnets. It's just that they come in strips, north, south, north, south, and so it's a little bit more difficult to notice easily. If you buy a bar magnet, it will usually be labeled. There will be a north end and a south end, and you will find that like magnetic poles repel, just as like charges repel, and opposite magnetic poles attract, just like opposite electric charges attract. Once again, it looks like action at a distance, and this is the beautiful part about buying those toys. If you hold two magnets together in the repulsion orientation, you can just visualize that magnetic field. It's the most kinesthetic experience of an action at a distance force that I can think of, a very nice feeling. You touch them, and you can really visualize that force field in between them.

If you start investigating magnets, you discover that the only thing that responds to magnets is other magnets or iron, which indeed is becoming a magnet when you hold the magnet close to it. A compass needle is just a little magnet, and compass needles will swing around and point so that you can use compass needles to detect magnetic fields, just as we use little charged up test objects like a balloon to measure electric fields.

You could map out the magnetic field in the room, and the easiest way to do this—it's kind of fun—would be to take a magnet, lay a piece of paper on top of it, and then go to a hardware store or maybe a scientific supply store and buy some iron filings. If you have somebody who works in a metal shop, you can find some metal iron filings. You want them to be very small, and light, and iron so that when they're in the presence of a magnetic field they become little compass needles. You just sprinkle them on the paper, and you'll see before your eyes the magnetic field lines appear because each little iron filing lines up with the field lines. Field, remember, represents the force right there on the object, and so all of the little filings line up. Because they're so light, if you jiggle the paper a little bit, they'll form these beautiful field line patterns just like the field line patterns that Faraday was drawing in the abstract. You can see them right there in front of your eyes. It's very manifestly physical, and that's a nice thing.

It's a little bit tougher to do this with electric fields because there's not a simple object that will line itself up like iron filings do with the magnetic field although if you are clever, you can come up with various materials that will visualize electric fields. Magnetic fields can be very strong, and that helps in this story because if you buy a bar magnet, there's a really intense field right next to it.

There are some of these commonalities between electric fields and magnetic fields but also some very clear differences. You cannot create a magnet by rubbing anything with anything. Most materials are not magnetic. It's only very special materials, and once you have a magnet, it tends to stay magnetic for an arbitrarily long period of time. Kitchen magnets don't wear out. Compass needles do not deflect in the presence of electric charge. If you charge the balloon on your sweater and hold it near that compass needle, there will be no impact. The magnet is neither attracted to nor repelled by the electric charge. You can stop right here. We have just proved without doubt that electricity and magnetism are completely different, unrelated forces of nature, because if magnets were doing their thing, if they were attracting and repelling because of some little build up of static charge inside of them, you could convince yourself with this one simple experiment that it's not static charge inside of a magnet. It's some different physical thing.

Now, I encourage you to think about lots of experiments because there are lots of different ways that you could convince yourself of this idea. Here's another one. Earth is a giant magnet, and you can tell because compass needles always point in the same direction. But, balloons don't all float in the same direction. There is no attraction of electric charges to magnets just as there is no attraction of magnets to static electric charges. It doesn't work either way. Here is another big difference. In fact, it's a huge difference between electricity and magnetism. You take that balloon, and charge it. It's all negative now, and I can hand it to you and you can walk away with that balloon. We have separated charges in the universe. We didn't create them, but we did separate them. Now you have an isolated negative charge. You could imagine, at least in principle, holding a single electron between your fingers, one isolated electric charge. You cannot do this with magnets. You might say, well, why not? I'll just buy a bar magnet, and I could try this experiment, and take a hacksaw and split it in half. I'll pull the two pieces apart, and wouldn't I have a magnetic north pole and a magnetic south pole? Well, I encourage you to try it, and what you'll discover is that you don't have a north pole and a south pole. You have two smaller magnets. You have a north/south and a north/south, and when you separate them, you still have two little smaller magnets. You didn't isolate a north pole. You simply created a new north/south pair. If you could isolate a magnetic pole all by itself, physicists have given this a name even though we've never found it in nature. It would be called a magnetic monopole—monopole, a single, magnetic pole. A magnetic monopole would be just like a single isolated electrical charge. We haven't found it, which means there's a big difference between electricity and magnetism. With electricity, you generate electric fields from an isolated electric charge, but you don't generate magnetism from an isolated magnetic charge or monopole. There's something quite different about them.

Here's the bottom line. If you have electric charges, they interact with other electric charges. They attract or repel. If you have a magnet, it interacts with other magnets. They attract or repel. If you have a mass, it interacts with other masses. That's gravity. It only attracts. These three forces seem to be living in their own little worlds. They deal with their own special materials. They seem to be completely distinct and unrelated except for the slight mathematical similarities, like the fact that electricity and gravity both decrease like one over distance squared. Magnets, it turns out, decrease at a different rate, and we'll have to come back and think a little bit about the origin of magnetic fields to make sense of that.

In the early 1800's, this was the state of physics knowledge about electricity, magnetism and gravity. We understand the basic principles. They seem to be separate and distinct things, and then in the early 1800's a scientist named Hans Oersted in Denmark was preparing for one of his classes when he was lecturing to the students about these new, esoteric, laboratory phenomenon of electricity, batteries and compass needles. He was going to do everything all at once up on the stage, so he had a battery and some wires. He could close a switch so that electric current would flow through the wires so he could talk about electrical phenomena, and he had a compass needle up there. What he discovered by accident, preparing for class, was that when he closed the electrical switch so that electrical current was flowing, the little compass needle twitched. This was a huge surprise to Oersted. Apparently, he showed it to his students, and they were relatively disinterested because it wasn't going to be on the test, but Oersted recognized that this was a fantastically interesting phenomenon because remember, up to this point, electricity was electricity. Charges talk to other charges. Magnets talk to other magnets, and there's no interconnection.

Now we see an interconnection. We see that flowing electric charges in the circuit do make a magnetic field. They attract a magnetic compass needle. This is a big change in the way people are thinking about electricity and magnetism. All of a sudden, you have this very simple experiment, totally easy to reproduce in other laboratories, and it starts generating an enormous amount of interest. Lots of scientists were excited by electricity. It seemed that it had a lot of promise in the early 1800's. There's this famous quote, which may be Apocryphal, in which a government minister spoke to Faraday and said, "Of what use is this electricity phenomenon?" And Faraday said, "I'm not sure, sir, but some day you will tax it." It's very insightful if it is a true story.

People are experimenting. They're mucking around. They're thinking about how to use this electricity thing and how to use this magnetism thing, so now you see this connection. You can close a switch over here— and mechanically, it's very easy to close a switch—and current flows, so now you can make a solid metal object, such as an iron bar, rotate. You can cause motion with electrical and magnetic phenomenon. Once you can make motion, now you're thinking engines and dynamos, all of these practical applications. What a wonderful idea you have at your fingertips. We really need to make sense of this because if you want to manipulate it, you really need to have some underlying picture of what's going on.

One of the reasons I love playing with magnets is that they feel like magic. I love that kinesthetic feeling, and making sense of this feeling and controlling it is part of the sort of philosophical drive, I think, for this period of time. It's partly practical and partly just this curiosity about what the heck is going on here. André Ampère, who I mentioned before—we've named the unit of electrical current after him—is a French mathematician, and he is interested in this experimental science. He begins to do some experiments, but largely he is doing theory. He is looking at data, and he begins to formulate a law of nature that explains magnetic fields, very much as Coulomb created a mathematical law of nature that explained static electricity.

Here is the bottom line of Mr. Ampere's law. Static electric charges, charges sitting still, create electric fields. If you think of the pattern, the field lines, they emanate from electric charges. They start at positive electric charges. They run away, and they end at negative electric charges or else they run off to infinity. Ampère discovers that magnets are quite different. You can look at the magnetic field lines, and let's think about Oersted's experiment, the one where you have a wire with current running through it. Let's have a long wire that goes from the floor up into the roof. This is a big experiment, and now let's have the wire run through a hole in the table, and you take your piece of paper and sprinkle some iron filings on it and ask the following—what will be the pattern of magnetic field produced when the current is running straight up the wire and you have this piece of paper around the wire? You might think by analogy to electricity that you would see radial magnetic field lines pointing away from the wire. That would be a natural, reasonable guess, and it's completely not what you see. What you would see in the laboratory would be circles running around the wire. The little iron filings show you the magnetic field. A little compass needle would verify that it is the direction of the magnetic force, and it runs in a circle around the wire. It doesn't begin or end anywhere. A circle has no starting point and no ending point. This is a huge difference between electricity and magnetism. Electric field lines begin and end on electric charges, but there is no physical analogy for electric charge in the world of magnets. There is no magnetic monopole.

Magnetic field lines are different, and when I look at a magnetic field line, sometimes I think of it as arrows. Arrows carry with them a little risk because you think of something flowing. Fields aren't flowing. The field is static. The charge is flowing, but it just creates a pattern in space. The magnetic field is just sitting there, and when you draw it, the shape of the field is circular. As you go farther and farther away from the wire, the magnetic field is weaker and weaker. Ampère, again, wrote down the mathematics so that we could predict the magnetic field generated from any current going in any direction of any strength.

This affect is very real. It's not that hard to reproduce. I would put it at the level of a good, hard, high school science fair project. If you try to go to the basement and see this for yourself, you have to realize there are a couple of difficulties. One is that you need pretty hefty currents, and hefty currents can be dangerous. They'll heat up wires. You might set things on fire. You'd probably need car batteries in order to generate the kind of current that you're interested in, so you'd want to use some care. There's another element of this experiment that makes it a little bit difficult, and that is that there's always the Earth's magnetic field. If you're creating your own magnetic field, you're always superposing the one you're making with the one that's already there. What's the result when you take two magnetic fields and you put them on top of each other? Well, it's back to Galileo's super-position principle. It's just as if you have two forces acting. If the two force arrows act in opposite directions, they cancel out. If they act in the same direction, they add up, and you have twice as big of a force. It's the same thing with magnetic fields. If you have a magnetic field from your current in the wire and you have the Earth's magnetic field, if you align things so that they're both pointing in the same direction at some given point in space, you'll have a doubly big magnetic field, but if you've aligned things so that they're in opposite directions, you won't have any magnetic field at all in that point in space. This is the other difficult thing about that experiment—making sure that you're separating the natural field that's already there in the room from the created field that you're trying to study.

One of the deep and crazy things that Oersted has discovered that Ampère is making mathematically rigorous, but seems weird to people, is that static electric charges don't make magnetic fields. They don't interact with compass needles, but moving electric charges do. As soon as you have a current flowing, then you have magnetism in the room. This is very weird. Let's think about a very mechanical analogy. Suppose you have a balloon, and you've rubbed it on your shirt so it's highly charged. It creates an electric field. Other electric charges will respond, but magnets will not notice it. There's nothing magnetic about it. There's no magnetic field in the room. Okay, I'm sitting in the room, and there's the balloon. I am insisting and quite confident that I have experimental evidence with my compass needle that there is no magnetic field in the room.

Now I want you to enter the room in a slightly difference reference frame. Remember, we've argued it's basic Newtonian idea that you can observe the world in any inertial reference frame you like. You can be sitting away from me and make your measurements. You can even be moving with a constant speed in a straight line with respect to me. So, you're in a little cart, and you're cruising into the room. You go past me, and you cruise out of the room. Let's say you're moving at a steady two meters per second. You're in this little cart, and in your reference frame, you would consider yourself to be at rest. This is a critical idea about reference frames. You're in the cart. Everybody thinks they're at rest, but what you see is you're at rest in a cart and you see a room slide by you at two meters per second. You see me slide beside you holding a compass needle. You see a balloon slide by you. Think about that balloon. You see an electric charge moving past you in your reference frame. A moving charge is an electric current. Electric current is nothing more than moving charges. You see a current in the room, and you say, huh, according to Mr. Ampère and Mr. Oersted, if there is a current, there will be a magnetic field. You pull out your compass needle, and surely enough, the compass needle changes its orientation as the balloon goes by because that's direct evidence that there is a magnetic field in the room.

You and I are standing right next to each other. You pass right by me, and you and I disagree about this fundamental statement. Is there or is there not a magnetic field in the room? This is a wild idea that is leading us to a much deeper understanding of the principle of relativity—first, Galileo's relativity, and later, Einstein's relativity. What we're saying is that, yes, electric fields are very real. You can see them, though they're a little bit tougher to see. You could sprinkle grass seeds that are electrically polarized and see them. You can see magnetic fields with a piece of paper and iron filings. They're definitely there in the room, and yet you and I can disagree about the strength of magnitude of the fields.

Now, that's kind of a wild idea. Then you just say, okay, I guess that's right. You and I disagree about lots of things. We're in different reference frames. We disagree about positions. We disagree about velocities. We apparently disagree about the strength of magnetic fields just as we disagree about the strength of meters per second. These are just numbers describing physics. What we agree on are the rules. We all agree that moving charges create magnetic fields. Stationary charges create electric static fields. Every law of physics that you and I come up with is in complete and total agreement. We discover that the mathematics Ampère developed, and further physicists beyond him are developing, applies to how you connect the values in the two different reference frames. It's pretty simple and straightforward, so this is a connection between electricity and magnetism that's really profound.

Think about this for a second. I'm in a room that only has electric fields. You're in a room that has magnetic fields, and yet we're both in the same room. It's the same space so electricity and magnetism seem to be connected to one another in an intimate way. It just depends on the observer whether you call it electricity or whether you call it magnetism. That is really the cool, deep idea that people are developing here. Electromagnetism is the name we're going to ultimately give to this one force of nature that just has two sides to it.

Let's go back to applications. The magnetic fields that are produced by an electric current are ordinary magnetic fields. They're exactly the same as a magnetic field from a kitchen magnet or one of those toy magnets, and it's really a universal law of nature that we're discovering here. If you have a certain current, you will create a certain magnetic field strength. A compass needle will twitch with a certain strength, and if you double the current, think of the superposition principle. It's just like having two charges flowing per second instead of one, and that's going to make two times as big a magnetic field, which is part of what Mr. Ampère is writing down in his mathematical formulas. If you double the current, you double the magnetic field. Now you're starting to think applications. What if I want a really strong magnet? What if I want to be able to lift a car up at a junkyard, a car that has some iron in it? I'm never going to be able to go to the beach in Greece and find a loadstone big enough to do that. What I'm going to do instead is I'm just going to flip a little switch that makes electricity flow and the flowing electricity is going to create this magnetic field. It's going to lift up the iron car, and I can turn it on and I can turn it off. Think of the technological power of this, because manipulating electric circuits is quite easy, and then you can do these mechanical things with it.

You can sit in your car and take the metal key, insert it in a slot, turn that metal key, and what is going on? Well, you've now reoriented the metal so that it's touching some other pieces of metal inside the jacket, and you've completed a circuit. Now electricity is flowing, and the electricity flows through a coil of wire called a solenoid, which makes a magnetic field that is strong enough to pull a little iron rod. This is how you start your car. In the old days you had to physically, mechanically turn a crank, and now you're letting the electricity turn on a magnet and the magnet does the pulling—the great manipulative power of these basic laws of physics.

If you want to understand magnetism, now you ask the question, what about the kitchen magnet? I was just arguing that magnets arise because of electric current. You can turn them on, and you can turn them off. What's the deal with the kitchen magnet or with the toy magnet? That's a nice story too. We can make sense of this, but to make sense of it accurately and completely, requires some quantum physics. The classical story is good enough. It's a nice, simple explanation. Let's think microscopically. We have some material object. It's made of atoms. Atoms have a positive, heavy nucleus and a little negative electron running around on the outside, a little electric charge running in a circle. That's a current, and the current has a definite direction to it. It's going around in a circle, and according to Mr. Ampère's law, that's going to create a little tiny magnetic field. It's not a battery that makes it happen. Any time you have a moving charge in nature anywhere—it's universal—any moving charge creates a magnetic field in a very well-defined way. Little atoms create little magnetic fields, which means that atoms are little tiny magnets. Now, imagine that you have one atom here and another atom right next to it, and maybe the electron is going clockwise in one of them and counterclockwise in the other. It's three dimensional, so the little magnets are totally randomly aligned. They're very close together, so if you have an up magnetic field from one atom, and a down magnetic field from a nearby atom, up and down cancel. It's the super-position principle applied to magnetic forces just like any other force. It's just Newtonian or Galilean super-position.

Ordinary materials aren't magnetic because the atoms are randomly aligned. All of the little micromagnets are randomly aligned. It requires a very special material, one in which the electric current of one atom couples or connects to its neighbor in such a way that the two magnets tend to line up. Iron crystals microscopically are such that the atoms tend to like to line up their magnetic fields. Now, how is it that you can create a magnet out of iron? Well, you rub an existing magnet against an iron bar, and as you do so you're creating a strong magnetic field that begins to line up all of the atoms because after all magnets like to line up with other magnets. Once it happens, you have a cascade because each atom's magnet talks to its neighbor, which talks to its neighbor, and there is this sweep through the iron as everybody lines up. Now you step back and look macroscopically, and you see a whole bunch of little itsy bitsy microscopic magnets that are all pointing the same way. You have a magnet in the laboratory, and once they're lined up, they tend to stay lined up. There's no electricity required. You don't have to have a battery to make a kitchen magnet stick to the refrigerator.

There is this natural or automatic way of creating a magnet or this artificial way of driving electric charges through electric circuits. You use electric fields to push the charges to make them go in a path, so there are these multiple ways of thinking about magnets that are really ultimately all the same story—moving charges.

Now, we've argued that static charges do not make magnetic fields, but moving charges do. We've also argued that static magnets do not make electric fields, but what about moving magnets? It's a natural question to ask. It seems like, gosh, if the universe was symmetrical, moving magnets should make electric fields. This is what Mr. Faraday investigated. One of the many wonderful experiments he did was to take a magnet and to move it rapidly, to jiggle it. Any kind of motion will work, and by the way, you should appreciate that these effects are fairly subtle. When I told the story about a moving balloon making a magnetic, the field that would be an extraordinarily weak magnetic field. If you move a very, very powerful magnet rapidly, you might just be able to notice this effect in the laboratory—this is Faraday's discovery. It's perhaps the most practically important applied physics discovery in the history of modern science, because if you can wiggle a magnet and create an electric field, think about what you can do. You can make electric fields. That means you can move charges around. That's what we want in our houses.

Where is the electric field coming from that pushes the electrons through your toaster oven, or through the lamp, or through your computer? Well, somewhere out of town, there is a power station, and it has a big giant magnet, a physical giant magnet. Somebody is turning it. It might be a windmill turning it. It might be a steam engine turning it, but there is a big magnet that's turning near a metal coil. The changing magnetic field is driving electrons, pushing them, all the way to your house. It's a direct physical connection, and it's this entire symmetrical story that electric and magnetic fields are indeed after all intimately connected to one another. One of them can create the other, not statically—you have to have motion. But, as soon as you have motion, moving charges make magnetic fields. Moving magnets make electric fields—beautiful symmetry. We're going to want to come back and tackle this symmetry. It was the work of James Clerk Maxwell to recognize how elegant and beautiful this story ultimately is.

Lecture Seventeen
Unification I—Maxwell's Equations

From a long view of the history of mankind— seen from, say, ten thousand years from now, there can be little doubt that the most significant event of the 19th century will be judged as Maxwell's discovery of the laws of electrodynamics. The American Civil War will pale into provincial insignificance in comparison with this important scientific event of the same decade.

—R. P. Feynman, *Lectures on Physics*, Vol. II

Scope:

In the last lecture, we saw that electricity and magnetism are different forces of nature, but they seem to be connected in certain ways. For example, a moving magnet produces an electrical field and vice versa—a moving electric charge produces a magnetic field. In this lecture and the next, we'll refine our understanding of these forces and their connection. The unification of electricity and magnetism is one of the grand intellectual achievements of classical physics. The person credited with this synthesis is James Clerk Maxwell, a Scottish physicist working in the mid-1800s, who organized the work of Ampère, Coulomb, Faraday, and others into four simple equations that constitute the "rules" of electricity and magnetism. In the end, Maxwell was able to summarize everything we know about electromagnetism in a set of four relations, two for static situations and two for time-varying situations. Together with Newton's laws, these relationships quantify all electric and magnetic phenomena. In this lecture, we'll bypass the mathematics of Maxwell's equations and try to understand the underlying essence of each one.

Outline

I. The first of Maxwell's equations is similar to Coulomb's law; it describes electrical fields arising from electric charges.

A. Maxwell combined Coulomb's work with that of a German mathematician named Carl Friedrich Gauss (1777–1855). *Gauss's law* involves looking at the outside of a region with an electrical field to deduce the nature of the sources inside. Think of encasing an electric charge in a bubble and observing the electrical field lines emanating from that source. If field lines are pointing outward in all directions, we can conclude that there must be a charge inside the bubble.

B. Gauss's law is intuitive: Electric charge is the source of electrical fields, which emanate from the source. The same is true in reverse: If we see electrical fields pointing outward in all directions, we know that they must be emanating from a source.

C. Gauss's law is both quantitative and qualitative: It tells us how strong the electrical field is and what pattern it produces. Gauss's law is also more robust than Coulomb's law. It accounts for multiple electric charges and for moving charges.

D. Gauss's law is universal, requiring a fundamental constant of nature (measured by Coulomb). This constant is necessary to determine how strong the electrical field is for a given amount of charge.

E. A dog sniffing around a barbeque grill can conclude that the smell of cooking food is emanating from the grill as a source. Gauss's law can be thought of in the same way.

II. The second of Maxwell's four equations is sometimes called *Gauss's law for magnetism*.

A. This is a sort of negative law. If we encase a magnetic field in a bubble, we will *never* see field lines emanating from a source in the middle. In other words, there are no point-like sources of pure magnetic field lines in the universe—no magnetic monopoles.

B. Even though this law is negative, it tells us something about the pattern for all magnetic fields in the universe: They never emanate from a point; they run in circles, never stopping or starting at points in space.

C. Physicists have sought to falsify Gauss's law for magnetism (by finding a magnetic monopole) for 150 years, without success.

III. The third of Maxwell's four equations is also called *Ampère's law*.

A. As mentioned in the last lecture, Ampère's law tells us that a current (flowing electric charges) generates a magnetic field. Ampère's law is a rigorous description of this phenomenon, a mathematical formula that can be used to calculate the strength and direction of the magnetic field, just as Gauss's law can be used to determine the strength and direction of the electrical field.

B. Again, Ampère's law requires a numerical constant of nature to characterize the connection between the amount of electric current flowing and the strength of the magnetic field.

IV. The fourth of Maxwell's equations arose from Faraday's work and is usually called *Faraday's law*.

A. As we mentioned in the last lecture, Faraday realized that a moving magnetic field generates an electrical field; in turn, electric charges will respond to this field.

B. Faraday's law is the most directly practical and widely used of all the equations described so far. Sometimes called the *law of electrical induction*, it accounts for how we produce electricity in our homes and how we convert electrical voltages (via transformers) for such devices as laptop computers and cell-phone chargers.

V. In looking at these equations, Maxwell noticed an aesthetic "hole," a lack of symmetry.

A. According to Ampère's law, flowing electric charges create magnetic fields. At the same time, according to Faraday's law, changing magnetic fields produce electrical fields. If that's true, why wouldn't changing electrical fields produce magnetic fields?

B. To resolve this problem, Maxwell took a leap and hypothesized an additional term in Ampère's law, designed to make it symmetrical with Faraday's law. In making this leap, Maxwell drew on the rich tapestry of data from existing experimental and theoretical physics and realized that his hypothesis would have to stand up to the tests of mathematical and physical consistency, consequences, and falsifiability.

C. It took many years for Maxwell's addition to Ampère's law to be directly tested through experiment, but once it was, numerous practical applications were realized.

D. We now have two ways to produce a magnetic field, but why don't we have two ways to produce an electrical field? The answer: There are no flowing magnetic charges—no magnetic monopoles—in the universe.

VI. Maxwell's four equations enable us to understand the patterns of electrical and magnetic fields in any circumstances. Let's summarize them.

A. Static charges generate static electrical fields that show a radial field pattern, but there is no analogous source for magnetic fields because there are no magnetic monopoles.

B. Moving electric charges generate circular magnetic fields; further, a moving electrical field also produces a magnetic field.

C. Finally, a moving magnetic field produces an electrical field.

VII. Maxwell's equations focus on fields, leading us to think of nature, as we do today, in terms of field theory.

A. These equations are enormously practical. They tell us how to design devices ranging from a cell-phone antenna to a toaster oven.

B. Further, all of Maxwell's equations tie in with Newton's laws and help us see fields as "real." The equations fit beautifully with the classical physics worldview, enabling local, deterministic, and quantitative predictions and explanations.

Essential Computer Sim:

Go to http://phet.colorado.edu and play with Faraday's Electromagnetic Lab (check out the Transformer and Generator tabs if you haven't already). Then look at Radio Waves and Electromagnetic Fields. Can you understand the various representations of an electrical field? Which of Maxwell's equations are involved?

Essential Reading:

Hewitt, review chapters 21–24 (it's all there).

Hobson, start of chapter 9.

Recommended Reading:

Cropper, chapter 12.

Gonick, start of chapter 23.

Questions to Consider:

1. If magnetic monopoles existed in nature, what changes would we have to make to Maxwell's equations? Which equations would be modified? Would there be a new constant of nature to measure?
2. I have argued that Faraday's law (which states that any change in the magnetic field through a coil will generate electric currents) has had enormous technological impact, specifically with regard to electrical generation and transformers. What other pieces of everyday technology make use of Faraday's law? (Think of devices in your home, in your car, at the airport…) The list is quite long; can you come up with a half-dozen?
3. We have said that electric currents generate magnetic fields. Is there a measurable magnetic field near, say, the cord leading to a lamp in your house? (If not, why not?) We have said that *changing* the electrical field over time also generates magnetic fields. Since the current in the lamp cord is AC (alternating current, flowing back and forth), would *that* generate a measurable magnetic field? Why or why not?
4. Static charges feel a force only from electrical fields. Moving charges feel forces from magnetic fields, as well. This fact forms the basis for electric motors: You use electrical fields (e.g., from a battery) to run current through a metal loop situated in a static magnetic field. The current (moving charges) feels a force from the magnetic field and is pushed—it turns. That's all there is to any electric motor. What happens if you take the exact same device but disconnect the battery? What happens if, with the battery disconnected, you physically rotate the metal loop in the presence of this magnetic field? What important device have you just created?

Lecture Seventeen—Transcript
Unification I—Maxwell's Equations

We've been talking about electricity and magnetism. These are two fundamental forces of nature that are definitely distinct from one another and distinct from gravity, and yet what we've seen in the last couple of lectures is that although they are different from one another, they're also connected to one another. Somehow electricity and magnetism seem to be related in ways that we need to understand more deeply. If you create an electric field by just holding up a charge, then it affects only other charges, and if you create a magnetic field by holding up a magnet, it affects only other magnets and not charges. However, if you wiggle the magnet, then you can affect the charge, and if the charges are flowing, then you can affect the magnet. There is some connection or crosstalk between the two.

In the end, what we'll be talking about today and in the next lecture is the synthesis of the ideas, the fundamental underpinning rules of electricity and magnetism. We'll see indeed how they connect together, how they weave together, and we will synthesize the theory of electromagnetism to form one of the most delicious, satisfying, and enormously practically important frameworks in all of classical physics. The person to whom we attribute this intellectual work is James Clerk Maxwell. Maxwell was a Scottish physicist who was working in the mid-1800s, the time of the American Civil War. He is a contemporary of Michael Faraday, and he's not a Newton. He's not coming up with radical new ideas about the universe from scratch. It was much more in the contemporary sense of standing on the shoulders of giants.

Mr. Maxwell knows about the work of Coulomb and his laws of static electricity. He knows about the work of Ampere and his formulas that explain how moving charges make magnetic fields. He knows about Faraday's work and how moving magnets can create electric fields, and Maxwell was able to take the work of these other people, organize it, think about it in a new mathematical way, and write down a very simple set of equations—the fundamental equations of electricity and magnetism, which today are called "Maxwell's Equations" in his honor. There are four of them, and that might be a little bit of a surprise to you because, after all, there are only two forces. You have one law of gravity that Isaac Newton came up with that we've been using. Why not one law for electricity and one law for magnets? That's a nice idea, and people tried to do that. Coulomb was coming up with a single formula for static electricity, but it is the interconnectedness of the two forces that requires us in the end to have four equations (or four relationships), two of them dealing with static situations: a charge sitting still, a magnet sitting still—and the other two for time-varying situations: moving charges, moving magnets.

Together with Newton's laws, these relationships are quantifying every electric and magnetic phenomenon that you can think of. They predict. They describe. They explain. It's just perfect classical physics. It fits in exactly with Newton's laws, and we're really zooming in on electric and magnetic forces. Once you know about electric and magnetic forces, then Newton takes over and says F = ma, and so now I know how things respond. This "how things respond" is really the practical side of electricity and magnetism where these classical equations are the bread and butter of contemporary physicists, engineers, and scientists of all disciplines.

Mr. Maxwell was a theorist. In the old days, back in the Newtonian era, people were usually both. They would do experiments. They would think about the theory. They would work on both parts together. As our knowledge deepened, there was a split, and some people focused more on the experimental work. Maxwell pretty much just sat in his office and did mathematical calculations. He was a remarkable mathematician, and part of the difficulty of this story is that if you just look at the original mathematics of Maxwell, it's really dense. It's hard stuff. What I want to do is think about the essence, the fundamental ideas, that these mathematical relationships are telling us. We're going to go through Maxwell's Equations today, look at what they say, try to make sense of them, see how they're telling us about electric and magnetic fields and recognize that fields are now the critical player in this story.

When we started, we were thinking about forces. There's an object over here. There's another object over there, and they interact by some action at a distance. Mr. Maxwell is asking us to think in a new way. He's building on Faraday's work. Remember, Faraday drew pictures of these field lines. You can draw a sketch of fields. A charge has field lines emanating from it. A magnet has magnetic field lines surrounding it, and when you look at the field, you're thinking about forces in a new way. You're thinking about forces as something local. For instance, if I have a big charge, it's the source of an electric field, and then I move far away. There's an electric field everywhere in space, and whether or not I put down a test charge, there is an E field out there. If I put down my little test charge, I know immediately what force it will feel. That's what electric field tells me. It tells me what force you would feel as soon as you put something there. Maxwell is focusing our attention on the fields, and he says these are the central players of physics. If we could understand where fields come from and how they interact one with the other, then we will really understand everything we need to know about electricity and magnetism because at this point it would just be Newton's laws. Plunk down a charge. There is a force on it, and it starts to accelerate. This is a shift of attention that's going to turn out to be extraordinarily productive. Nowadays, physicists pretty much always think about fields and field theories because that's really where the action is.

Let's walk through the four Maxwell's Equations, and we'll discover that we know most of them already. They were developed historically, and Maxwell is just reformulating them and rethinking about them. The first one is sort of Coulomb's law. It's the formula that tells us about how electric charges produce electric fields. It's sometimes called Maxwell's first equation, and sometimes we call it Coulomb's law. I'm going to call it Gauss' law. Carl Friedrich Gauss was a German mathematician. He did a little bit of physics, but was mostly thinking about the mathematics of fields in a way that turned out to be very, very applicable to electric charges. Maxwell pulled together Coulomb's experimental ideas with Gauss' mathematical ideas to formulate Gauss' law.

In essence, Gauss' law is just saying if you have an electric charge, then there will be electric field lines emanating from it. That's really the content of Gauss' law, but the way it's formulated is in kind of a clever and interesting way. Gauss' law, as thought up by Maxwell, suggests the following crazy idea. Draw a sphere or a bubble. It doesn't have to be a perfect sphere. It just has to be some closed surface. It's imaginary. You don't really have to build it, but just imagine that you have some bubble in space. Start walking around and just look at the surface of the bubble. Don't look inside. Just look right where you are. Stay local. This is one of the important ideas of field theories, and look at the electric field everywhere on the surface. If it's pointing out here, and then you walk a little further, and it's also pointing out there, and you keep walking around and around the sphere, and it's pointing out everywhere, Gauss' law concludes there must be a charge inside. It makes sense. If you picture the electric field emanating from a charge, it points out in all directions. It's a radial field, and so if you were to walk around it, everywhere you go the electric field would always be pointing outwards.

The point is that you don't have to look inside if you don't want to. You can deduce what's inside by just noticing the patterns around the sphere in space. This is a novel idea. It's a way of formulating the mathematics that's rather different, and it tells us a couple of things. If you work out the mathematics, you discover that it conclusively derives what the electric field from a point charge would be: positive or negative, or if you have a bunch of charges, or if you have some charges floating around and moving. Gauss' law works in all of these cases, and so it's a little bit deeper and a little bit more robust than Coulomb's law. It's a universal law. If you want to make a measurement of the strength of the electric field caused by a charge, somebody has to tell you a constant of nature. In fact, it's Coulomb who tells us this constant of nature. We need to know for a given quantitative amount of charge inside how strong the electric field is going to be at some point outside. That constant of nature was derived in an experiment by Mr. Coulomb, and you only need to do it once. If you do it again, you're just trying to measure it more accurately, but that one constant of nature will describe electric fields in any circumstance at any place or time in the universe, no matter what's going on. It's a very robust, universal idea.

Gauss' law tells you about the pattern of electric fields. It says that they radiate away from positive charges. They radiate towards negative charges. Now, I use that word "radiate" in a very informal sense. There's no motion here. It's just a static electric field. It's just when you look at the picture, you see these arrows pointing out like rays from the sun. Gauss' law can be mathematically formidable, but I would argue that it's also deeply intuitive. I think my dog knows Gauss' law, and as evidence of this I have this barbecue on the back porch, and I put some stinky hot dogs in there. Then I close the barbecue up so you can't see the source of stink. My dog is walking. She's a few feet away from the barbecue, and she smells this stink flow that's emanating away from the barbecue. Then she walks a little bit further in a circle around the barbecue, and it's still flowing outwards. She keeps going all the way around the barbecue, and everywhere around the barbecue the flow of smell is always outward. What would my dog conclude? She is quite smart. There has to be a source of stink in the middle, and that's really Gauss' law. If you see electric field flowing outward, there must be a source in the middle. If you see electric field coming in and then going out on the other side, then you know there's not a source of electric charge inside your sphere. That's Gauss' law. It's the first of Maxwell's four equations.

The second of Maxwell's four equations doesn't really have a good name. Some people call it Gauss' law for magnetism. It's sort of an equivalent law, but there's a funny deal with magnetism. Remember we argued last time there is no such thing as a magnetic charge or a magnetic monopole. Gauss' law for magnetism is kind of a negative law. It basically says that if you walk around in this imaginary sphere around any point in space, you will never, ever find a magnetic field configuration that always points outwards or always points inwards—because if it did, it would have to end inside the sphere on a charge. However, there are no magnetic monopoles. It's kind of a negative law, but it's still quite productive. It tells us something about the pattern of magnetic fields in the universe. If you look at them, they always form circles or loops. They never start or stop at any particular point in space, and that's a very important idea. If you want to visualize fields, these equations are telling us both quantitatively and qualitatively what fields look like and how they behave. There is no new constant of nature required because we have a zero here. You don't need to know how strong the magnetic field would be if there was a magnetic monopole because there are no magnetic monopoles. This law of nature is like all good laws of physics—easily and deliciously falsifiable. All that you have to do is find one magnetic monopole, and Maxwell would be wrong, and you would have a Nobel Prize. It's a big deal. Physicists are always looking for cracks and breaks in the existing theoretical framework, even something as cherished as Maxwell's Equations, something as profoundly useful. We're always looking. Could this break down? Is there something else going on in the universe? Did we misunderstand? For 150 years, people have been looking for magnetic monopoles; we have never found one, and there is good reason to suspect that we might not. The equations that I'm in the process of describing are quite complete, and they describe every phenomenon—many, many consequences—by combining these equations together. There are many observable, experimentally verifiable consequences that flow from the equations and from the postulate that there are no magnetic monopoles. So, you never know, but at the moment, the law of nature seems to be that they're just not there.

The third of the four Maxwell's Equations goes by the name Ampere's law. We already talked about Mr. Ampere and his careful analysis of the accidental discovery of Mr. Oersted. Remember that Oersted was doing an electric demo for his students and noticed that if you run electric current through a wire, you generate a magnetic field. Ampere really worked hard to try to make this rigorous and mathematical. How much of a magnetic field and how strong of a magnetic field is it? We need a new constant of nature for this law. You have to go out and measure it. If you have one amp flowing, one coulomb every second in the wire, and you go one meter away, how strong would that magnetic field be? You have to do one experiment just like Coulomb had to do one experiment. This is a new thing because you're measuring magnetism rather than electricity. It's the second fundamental constant to appear in Maxwell's Equations, and it turns out that's the only other constant that you need in Maxwell's Equations. These two constants of nature are not brought out from the theory. They have to be obtained from experiment, but once you have them, they're absolutely universal, and you can predict any other phenomenon in the world. You use your cat fur, and iron balls, and compass needles, and batteries—and in principle, that's all you need.

Once again, this is a deliciously falsifiable law of nature. I am telling you exactly in direction and magnitude how the magnetic field is going to look no matter what the current does. You can bend the wire, make it go in a circle or in any shape you like, and this equation will tell you what the magnetic field looks like. Every time you turn on your car and every time you turn on an appliance, you are testing Ampere's law because if there's a little magnet involved and you're producing a little electromagnet, the engineer who designed that circuit is counting quantitatively on Ampere's law working exactly as it was stated back by Maxwell in the 1860s. All of these laws have been verified time and time again, and by now they are so intimately involved in all of our technology that they are really deeply established laws of nature.

The fourth and last of Maxwell's four equations is usually called Faraday's law. Remember that Michael Faraday was doing lots of investigations. He is the fellow who gets us thinking about fields, and he's also investigating the properties of these fields. Remember, he's the one who noticed that if you take a magnet and you wiggle it, you will produce an electric field. Think about what that means. You're moving a magnet. It's just like a kitchen magnet or a bar magnet, any kind of magnet, and if the magnet is wiggling, that means the magnetic field that it produces is changing with time. Then you produce an electric field. It's pretty whacky because that means that if you put a charge there, a charge that doesn't notice or care about the static magnet, it now does react because an electric field means electric charges are getting pushed around. That's what electric field says. If you put a charge there, it will be accelerated.

So, this is a way of accelerating electric charges. It's surely the most practically important of all of the four Maxwell's Equations. This is the equation that tells us how to generate electricity at the power plant. You go out there and you take some big magnet, a physical magnet, and you rotate it. It might be rotated because of a windmill that's turning or because of a steam turbine that's running it in a big circle, but you just move the magnet and move it nearby some metal. If you're moving the magnet, you're creating electric fields. If you put an electric field on a wire, there are electric charges in that wire, and they're free to move—and so they will, and that wire connects all the way to your house. So, the electrons that are being moved in your house are being pushed by this spinning magnet using Faraday's law.

Faraday's law is sometimes called "the law of electrical induction" because we're inducing electrical flow by using magnets. When you take one of those little black boxes and plug it in the wall and connect it to your laptop or connect it to your cell phone, you're also using Faraday's law of induction. What is happening there is you have high voltage, 120 volts, at the wall, but your laptop or phone only wants a few volts. To change the voltage, you can make use of Faraday's law. The idea here is to remember that the magnet at the power station is turning in a circle. So, the magnet field is getting stronger and then weaker, stronger and then weaker, which means that the electric current that flows goes one way and then the other, one way and then the other. It's alternating current, which means that at your house this 120 volts causes current to flow first one way and then the other. If you make that go through a wire, in your house you can create a little magnetic field. That's Ampere's law, and then that little magnetic field is changing with time. Faraday's law says that is going to produce a little electric field, which will drive the electrons into your cell phone or your laptop. It's all a simple interconnected story, although it takes a lot of practice to really work out the details and understand the practical workings of these equations. The big idea is just what we've been talking about.

Maxwell writes down these four equations, and he's staring at them. He is able to write them down in a spectacularly compact form. They fit on a t-shirt, and that's a good sign that you have a fundamental theory of nature. He is staring at the equations, and he realizes that there is a problem. There is something aesthetically unappealing. It's Ampere's law that's really sort of bugging him because Ampere's law says if you have flowing physical electric charges, then you create a magnetic field. He says, okay, that's one way to create a magnetic field, but now look at Faraday's law. Faraday's law says that if you wiggle the magnetic field, you can produce an electric field. You don't need any charges. You don't need any magnets. You just need to change the magnetic field itself, and that changes or produces an electric field. It seems like if a changing magnetic field can produce an electric field, it should also work the other way around. There is this symmetry argument that a changing electric field should be able to produce a magnetic field. If it works one way, why doesn't it work the other way?

So, Maxwell takes a leap. He looks at the equations. He thinks about their mathematical form, and he adds a new term to Ampere's law. It's Maxwell's addition to Ampere's law, and it's a new term that says, okay, one way to make a magnetic field is to have flowing physical electric charges. The other way, the new way, to make a magnetic field is to have changing electric fields. It's a fundamentally different kind of way of producing a field. This is a risky business. You sort of get uncomfortable when somebody makes up an equation out of thin air that is not based on direct experiment. This is not something that happens all that often, although you could argue that many of the great, great discoveries in physics have been of this nature.

Now, I've criticized the Greek philosophers for just making up ideas about nature. They sit around. They think about symmetry or beauty, and they say, "I believe the world should be this way." Then they argue with one another about how the world should be. Maxwell is really not acting in that way. Maxwell has come up with a scientific hypothesis. Is this new term physically correct? Is it describing nature? He behaves just as a contemporary physicist should. He starts calculating. He asks himself does this new equation first of all agree with every experiment ever done in all of history? It is essential that you be scientifically honest. You have to make sure that your new idea doesn't violate any knowledge that you already have about the world, and you have to ask what further consequences it would have. How could we falsify this new idea? What real life experiment could we do to test my new, essentially crazy idea that there's this extra piece, this extra way of creating fields.

Maxwell worked very hard on the mathematics. He spent many, many years investigating the consequences, thinking about experiments, and thinking about practical consequences, and he really took it on as his responsibility because he recognized he was already living in a world of classical physics, which had started with Isaac Newton, and Galileo before him, and Copernicus before him. By this point, with the work of all of these other physicists we've been talking about: Coulomb, and Faraday, and Ampere, you can't just arbitrarily make stuff up or tweak it. You have this tapestry in front of you, and everything is fitting together beautifully. What he did was he just discovered a missing thread, and when he put the new thread in, it matched and fit perfectly, elegantly, and beautifully. It's a little bit hard to describe without looking at the mathematics just how you could look at an equation and say, "Ah, that's lovely." The idea that we're discussing really, I think, carries this with it.

The applications of this new added law turned out to take a while. It took many years before Maxwell's addition could be directly experimentally tested, and we'll talk about that in the next lecture. It came, and once it came, they began to come in droves. Ultimately, Maxwell's Equations became the fundamental equations of classical electromagnetism, which we still use today.

You might have one question if you look back at the story that we've told so far, and it's a slight puzzle, but we can resolve it. Faraday's law says that if you change the magnetic field, then you produce an electric field. That's Faraday's law. That's this practical way that we produce electricity in the laboratory, and you might say, "Hmm, I'm going to play the Maxwell game and look at symmetry, and say there should be another way to produce electric fields, another way besides changing magnetic fields and besides having static electric charges, which would be to have flowing magnetic charges. After all, flowing electric charges make a magnetic field. Why shouldn't flowing magnetic charges make an electric field?" They would if there were any, but remember the second law of Maxwell is there are no magnetic charges—there are no magnetic monopoles in the universe. There is no need to add this new piece to Faraday's law because we've already argued that there are no magnetic monopoles. The story is really very elegant and very complete. The four formulas all tie in one with the other, and they tell you about electric and magnetic fields and their patterns in space.

Let me remind you of the four stories. Static charges create electric fields that emanate outwards from them. There is no analogous magnetic pattern. There are no magnetic monopoles to create radial magnetic fields. They just don't exist. The third of the four said that if you have flowing electric charges, physical objects, you will get magnetic fields that run in circles around it. That's one way to make a magnetic field, and the other way, which is Maxwell's addition, is to have any mechanism that changes the electric field. Any other way you can think of to make electric fields change at some point in space will also produce a magnetic field. The fourth and last one says if you want to create an electric field, there's another way besides having static charges. It's Faraday's way of wiggling the magnetic field. Wiggle real-life magnets if you like, and anything you can think of that can make the magnetic field change over time that will produce a particular pattern, which is a very well-defined pattern and a very well-defined strength.

We call them Maxwell's Equations even though he really didn't discover any of them except that one extra term, but James Clerk Maxwell did something really spectacular for us. He pulled it all together, 100 years' worth of hard work, and if you look at the old papers, the mathematics was complicated and hard to write down. People hadn't thought of elegant, simple ways of writing these ideas down, and this is one of Maxwell's grand accomplishments. He is also forcing us or encouraging us to think about electricity and magnetism as a field theory rather than action at a distance. He is helping to evolve Newtonian physics. It is still definitely classical physics, and it's enormously practical. If you want to design a cell phone antenna, you go to Maxwell's Equations. Maxwell's Equations will tell you how the electrons will respond to an electric field. Do you want to build a better microwave oven? It's Maxwell's Equations. Maxwell's Equations tell you how the electrons will move in the oven. As you wiggle them here, what will they do to the electrons in the water molecules in the food that you're trying to heat up? Any situation of any kind—high-tech, low-tech, a light bulb, or a toaster oven—it's all Maxwell's Equations, and ultimately it's Maxwell's Equations and Newton's laws because Maxwell's Equations tell you about the fields. However, the fields are these abstract quantities, which then tell you about how real-life objects, little magnets and little electric charges, will respond. How will they move? It is F=ma. They will respond to the electric and magnetic force in the usual way.

Maxwell is making us think about electricity and magnetism as belonging to the Newtonian picture. So, if you think about energy, energy and energy conservation better continue to hold. It's part of the framework or the tapestry that he's working with. He thinks explicitly about this, and he realizes that the mathematics tells him how it is that an electric field could, in principle, carry energy or momentum, which is yet another way that we really begin to think of these electric fields as something real. These magnetic fields are out there. They're a property of space. They live out there. You can't see them directly. You only notice them if you put a charge there or if you put a magnet there, but they're there anyway. They allow us to think of physics in a local way. It's completely deterministic. It's sort of the epitome of classical physics.

Lecture Eighteen

Unification II—Electromagnetism and Light

And God said:

$$\nabla \cdot E = \frac{\rho}{\varepsilon_0}$$

$$\nabla \cdot B = 0$$

$$\nabla \times B = \mu_0 J + \mu_0 \varepsilon_0 \frac{\partial E}{\partial t}$$

$$\nabla \times E = -\frac{\partial B}{\partial t}$$

and there was light.

—Maxwell's Equations on a t-shirt popular at MIT when I was an undergraduate

Scope:

Maxwell's equations synthesize and describe every aspect of classical electromagnetism, from lightning bolts, to electric circuits, to kitchen magnets. But Maxwell made another observation that went far beyond his original equations: He discovered they predicted a "new" phenomenon, an electromagnetic traveling wave, ultimately recognized to be light. All of optics, the remaining great branch of physics, was suddenly completely and deeply unified with electric and magnetic phenomena. Maxwell had provided a grand synthesis of all known fundamental forces of that era (*besides* gravity), allowing us to make sense of the spectrum of radiation and an enormous span of physics, as well as setting a compelling tone for ongoing physics research.

Outline

I. The most common motion in the universe is oscillation. The Earth, an atom, and a pendulum all oscillate. Maxwell asked: How would an electric charge behave if it were moving in this most common way?

A. A static electric charge (an electron) generates an electrical field with straight lines pointing, in this case, in toward the charge. You might think of these lines as similar to the gravitational field lines pointing in toward a star.

B. As the charge moves back and forth, the electrical field lines must move also in order to constantly point toward the charge.

C. A moving electrical field, according to Maxwell's "extra term," will produce a magnetic field; the resulting magnetic field will oscillate.

D. According to Faraday's law, a changing magnetic field will produce an electrical field.

E. The original source in this system was a charge, which created an electrical field, which in turn, created a magnetic field, which in turn, created an electrical field, and so on.

1. We can visualize this phenomenon by thinking of a pebble dropped into a pond. The pebble (the original charge) starts a disturbance (a wave) in the water that is self-propagating.
2. In Maxwell's case, there is no water or other medium. The ripples from the original charge are fields in otherwise empty space.

II. These fields exist, but we have to think about them mathematically.

A. We can find evidence for the existence of electrical and magnetic fields. Can we find similar evidence for the self-propagating phenomenon that Maxwell discovered?

B. We already know that we can detect the oscillation of an electrical field by looking for a similar oscillating response in a test charge.

C. Maxwell found that he could calculate the speed of propagation of an electromagnetic wave. The wave travels at a speed that is dependent on the two constants of nature from Maxwell's original equations. In fact, the answer turns out to be 300 million meters, or 186,000 miles, per second—precisely the speed of light!

D. This result implies that the electromagnetic disturbance we've been talking about is light; light is nothing more than a traveling electromagnetic wave.

III. Maxwell's discovery unified electricity, magnetism, and light.

A. Think about a light bulb. When the filament gets hot, the electrons inside begin to move, producing an electromagnetic wave that travels outward. The electrons in your retina respond to this wave as it passes, because an electrical field always moves charges around. The moving charges in your retina, in turn, send electrical signals to your brain. As we know, our brains respond only to a certain narrow range of frequencies of these oscillations in the retina, and we "see the light."

B. At this point, we understand the nature of light and optics and that the electromagnetic wave of Maxwell's equations is not exotic at all.

IV. Maxwell published his work over several years around 1860, but it took 20 years for the scientific community to accept this revolutionary synthesis.

A. Experimental verification came with the work of a young German physicist, Heinrich Hertz (1857–1894). Hertz built an electric circuit called an *oscillator*, designed to allow current to flow back and forth. He then built another oscillator, similar to the first one but with no power supply. According to Maxwell, the electromagnetic wave from the first oscillator should spread across the room at the speed of light and cause the second oscillator to respond—and indeed it did.

B. In effect, Hertz had built a radio. On one side of the room was an antenna broadcasting a signal, and on the other side of the room was an antenna receiving the signal.

C. Again, picture a moving electric charge at one point in space, which creates a ripple of electrical and magnetic fields that causes other charges, at another point in space, to move.

V. Maxwell's work opened up a new branch of physics—physical optics—that allowed scientists to think about optics in a new way.

A. Instead of exploring the path of light rays through a prism or how a telescope might focus light, scientists could now think about the interaction of light (electromagnetic waves) with matter.

B. Light had been studied since well before Newton; in fact, Newton's career as a physicist began with his experiments into the nature of color. He believed that light was "corpuscular," that is, made of particles, but the technology was not available to either prove or disprove this theory.

C. In 1800, Thomas Young (1773–1829), conducted an experiment that convincingly proved that light is a wave phenomenon. For 60 years after Young, physicists wondered: If light is a wave, what material thing is "waving"? As we've said, Maxwell answered this question: Light is electrical and magnetic fields oscillating in empty space.

VI. Anything we want to know about light should, in principle, arise from Maxwell's equations, including its origin, propagation, and interactions with matter. We should be able to understand lenses, rainbows, prisms, diffraction, and many other phenomena.

A. Maxwell's equations tell us that light carries energy. A moving electric charge is experiencing a force and, thus, accelerates over some distance; as we know, force multiplied by distance equals work, and work is a transfer of energy. Where does the energy go in this case? It spreads out in the electromagnetic wave.

B. The speed at which this wave propagates, 186,000 miles per second, is independent of any details of the motion.

1. Red light and blue light don't differ in their fundamental properties, but blue light has a higher oscillation frequency.
2. An even higher oscillation frequency than that associated with blue light won't be perceived by the human brain; this is *ultraviolet radiation*. Still higher oscillation frequencies result in other kinds of electromagnetic radiation, such as X-rays or gamma rays.
3. A slower oscillation frequency than that associated with red light results in *infrared radiation*. This is another form of light, and we can build night-vision cameras that detect this form of radiation.
4. In honor of Heinrich Hertz, we measure frequencies in units of cycles per second, now called *hertz*. The wall

plug in your house is 60 Hz; your eye responds to about 1 million billion Hz.

5. Beyond the infrared frequency is microwave radiation, and at a lower frequency still are radio waves.

C. Maxwell left a legacy of the unification of electricity, magnetism, and light, and an explosion of new ideas and applications that took off from his four equations.

Essential Computer Sim:

Go to http://phet.colorado.edu and play with Radio Waves and Electromagnetic Fields; Microwaves and Blackbody Spectrum are both worth exploring, too.

Essential Reading:

Hewitt, chapter 25.

Hobson, chapter 9.

Recommended Reading:

Gonick, end of chapter 23 and chapter 24.

Questions to Consider:

1. Why is Maxwell's change (addition) to Ampère's law needed in order to result in electromagnetic waves?
2. Is sound an electromagnetic wave? Why or why not? What are the similarities? What are the differences?
3. Suppose you had goggles that allowed you to see infrared radiation in much the same way you currently see visible light. What would the room you are in "look like"? In particular, what would be bright and what would be dim?
4. When astronomers see a distant supernova from another galaxy, they see a sudden increase in brightness of all colors of the spectrum (as well as radio signal, X-ray signal, and any other part of the electromagnetic spectrum they might be able to measure), all at the same time. How is this evidence that the speed of light is independent of frequency?
5. The Sun emits most of its energy in the form of electromagnetic waves, and most of that energy is found only in the near visible spectrum (from red to violet), with a peak in energy flow around

yellow. There is relatively little energy emitted in the infrared range (or beyond) or the ultraviolet range (or beyond). Our eyes, of course, are sensitive to just this same narrow band of frequencies that the Sun emits. Is this a remarkable coincidence, or can you think of a reason for it?

Lecture Eighteen—Transcript

Unification II—Electromagnetism and Light

Maxwell's Equations synthesize and describe every aspect of classical electromagnetism, all of it, any phenomenon that you can think about—lightning bolts, electric circuits, kitchen magnets, compass needles—just keep on thinking about electric or magnetic things, and they're all described by this simple set of four equations. We use them all the time, but Maxwell went beyond these equations. He looked at them, and it's one of those delicious situations where you've created these equations to describe a certain set of experiments, and then you realize that there's something new, something that you hadn't even thought about or imagined, and yet these equations describe it beautifully.

Maxwell found this by thinking about the most common motion in the universe. Ask yourself this question—what are random objects anywhere in the universe doing? There are a couple of different kinds of motion that you might find. Objects move in straight lines. Objects bounce and continue to move in straight lines, but the most common is in oscillation, something going in a circle like the Earth around the sun. If you looked at the Earth going around the sun from the side, you would just see it wiggling back and forth, back and forth. An atom in a crystal just wiggles back and forth. A pendulum wiggles back and forth. This is far and away the most common motion of objects in the world. It always happens if an object has a home spot, an equilibrium point, you perturb it, and it goes back and drifts on through. It oscillates back and forth.

It was a natural thing for Maxwell to ask, what would an electric charge do if it were moving in this most common kind of motion in the universe? What would an electric charge do if it were to wiggle back and forth? Maxwell's Equations should tell you the answer because they describe electric and magnetic fields in any situation. Let's think about it.

Let's think of an electron. If it were sitting still, there would be this electric field pattern that consists of nice, straight lines radiating in toward the electric charge. You could think of them as the gravitational field lines pointing toward a star, a nice simple pattern. Now, you move that charge forward and backward. At any point in time, the electric field lines are going to be described by Gauss's

Law. They have to point toward the charge. The charges are the source of electric field lines, so as the charge moves back and forth, the electric field lines have to wiggle in the space near and around the electric charge. Now you have a wiggling electric field, and Maxwell says, okay, I know about this. That was my extra term that I added for other reasons, just for symmetry reasons, to the set of equations. But the extra term very specifically says if an electric field at some point in space changes with time, it will produce a magnetic field. A magnetic field will appear out of nowhere, and it will appear in very specific places and in a very specific pattern. He works through the equations, and he realizes that as the electric charge goes up and down, so the electric fields go first up, and then down, and then up, and then down. The magnetic field that you produce goes first in, and then out, and then in, and then out. That's what the equations tell you.

You've created this magnetic field. It's somewhere nearby but not located at the charge, and now you say, wait a minute, I have a magnetic field and it's over here. It's bigger, and then smaller, and bigger, and smaller. It changes all of the time, and Faraday's Law, the fourth of Maxwell's Equations, says that at any time, for any reason, if the magnetic field changes with time, you produce a new electric field. There's this crazy thing going on. We started with the charge. That was the original source, but then it made an electric field, which created a magnetic field, which created a new electric field, which created a new magnetic field. It's this dance of electric and magnetic fields. Each one is produced, and because it wiggles, it produces a new one. The dance doesn't stop. It keeps on going. It's a self-propagating wiggle of electric and magnetic fields. It's a disturbance of electric and magnetic fields, and remember, this is in empty space. What is clear here is we're creating magnetic fields, and there are no magnets around. There is only one starting object and no other objects anywhere else.

I could visualize this a little bit like thinking of a pebble dropping into a pond. Imagine that you drop the pebble. That's the source. That's our moving charge, and what does it do? Well, it disturbs the water. It pushes water down, and then the water goes back up again. The water starts to wiggle, but if the water at one place wiggles up and down, water molecules feel the presence of nearby water molecules. If this one goes up, it pulls its neighbor up. Its neighbor starts to wiggle up and down, which makes its neighbor wiggle up and down, which makes its neighbor wiggle up and down. That's what you have when you drop the pebble into the pond. You see this ripple that spreads outward in a nice, elegant circular path heading outward. It's a disturbance that is self-propagating. It's a wave that heads out, and this is exactly what Maxwell sees. It's a self-propagating ripple except there's no medium there. There's no water or material object. There are no other charges. It's quite remarkable.

I've been arguing that fields have a separate reality. They exist in the world. You can think of them as existing at this point in space, but they're also abstract. You can't touch them exactly. You just have to think about them mathematically. This electric field and magnetic field that we produce is definitely real. There should be some evidence of it. It's going to carry energy because Maxwell has been thinking about the propagation of energy. Now he's a little bit worried because you had better not violate conservation of energy. You might worry that this dance spreads out and out and creates energy out of nothing, but, no, the equations are very clear about this. It's just like the ripple on the pond. It is weaker and weaker as you are farther away, but it doesn't ever stop. There is no friction in the electromagnetic case so although it is weaker and weaker, it never completely dies away.

In deep, empty space if you were to take an electric charge and wiggle it, Maxwell has realized that a new thing, a new physical event would occur, and he has to wonder if this is real. Have I discovered a new thing in nature? Is it a novel, undreamed of activity that electric charges do, and if so, could I really make it happen and could I detect it? How would I know that this was happens? Well, think about that. It's not so hard. If there is such a ripple in space, if the electric field somewhere far away from the original charge first points up and then points down, well, you know how to detect an electric field. Electric fields just mean that if there was a charge there, then the charge would feel a force. That's what electric fields are, so this electric charge that you put somewhere else goes go up and down. It bobs up and down, and that would be very easy to see. You would see an electrical response somewhere physically disconnected. What would you call this? It's an electric and magnetic phenomenon. It's an electromagnetic wave. Now, Maxwell is sitting as his desk—can just visualize this—and he's doing calculations. What would it look like? What would the pattern be? How fast would the ripple propagate? That's an important question, and the equations tell you. You have an electric field that's changing with time, and the equations tell you precisely how that time evolution creates a magnetic field and vice versa. It's all there in the mathematics, and it turns out that the wave that you create will travel at a speed that depends on those two constants of nature that appeared in Maxwell's Equations. Remember the ones that we had by rubbing cat fur and using compass needles? They tell us these constants of nature, and they combine in a fairly simple way. You multiply them and take the square root. This tells you the speed at which this electromagnetic disturbance, this wave, propagates.

I can just visualize Maxwell sitting with his slide rule, and he multiplies out the numbers. The answer comes out to be in the contemporary metric system 300 million meters per second. That's really fast, 300 million meters. That's 186,000 miles in one second. This wave goes from the Earth all the way out to the moon distance in one second. This is crazy. I wonder now if we could ever observe such a thing, and Maxwell must have been thinking about this number and realized, wait a minute, every physicist in the 1800's knows this number. Three hundred million meters per second is the speed of light. Is that a coincidence? Could that possibly be a spectacular numerical coincidence? It's this amazing revelation. I can't even imagine how it would have felt to recognize that this electromagnetic disturbance is light. That's what light is. It's just a traveling electromagnetic wave. All of the sudden, in one flash, Maxwell has realized that everything we have known, and thought, and understood about light now is connected. Light is. It's not a new thing. It's not some exotic phenomenon that we have to go and study and check out. We already know everything about this thing. We know about electromagnetic waves. It's just light. All of the sudden not only have we unified electricity with magnetism and made it very simple and workable, but also we've unified with light, optics, and the study of everything that has to do with those things.

Think about a light bulb. You heat it up, and when you heat things up, the little pieces inside start to jiggle fast. The electrons inside the filament jiggle back and forth, and if they jiggle fast enough, they're going to produce this electromagnetic wave. It doesn't even matter how fast they jiggle. They're always going to produce an electromagnetic wave, which travels outward in a spherically expanding path according to these equations, just like the light from a light bulb expands out in an expanding spherical path.

At some point, this electromagnetic wave you produce will reach your eyeball, and at the back of your eyeball, there are electrons. If there is this wiggling electric field pushing on the electrons in your retina, then they're going to go up and down, and that makes an electrical signal that goes to your brain. You go, ah-ha, I see light. Now, your brain only responds to a very narrow range of wiggling frequency in the retina so you only see a certain range of frequencies of oscillation, and that's why the light bulb has to go up to a certain temperature before you can see it. Even before you can see it, these exotic electromagnetic waves are not exotic at all, but are emanating from the light bulb. There are other electromagnetic waves coming out all simultaneously just different wiggle rates. That's the only difference between the different waves that are coming out.

At this point, we really understand the nature of light and all of optics. You start to think about all of the consequences. All of the things that you can now tackle quantitatively, rigorously, by going back to Maxwell's Equations have just expanded exponentially. It seems like a profound philosophical revolution that you have this new unification going on, but really, it has nothing to do with philosophy because it's just practical, painstaking calculations at this point. Maxwell's Equations are not easy to solve, but they are always solvable, and when you do, you can make quantitative predictions about anything that's involved.

Maxwell published this work in around 1860. There were a couple of works in a few-year period, and it took 20 years for the community to really begin to understand and believe that this was a revolution. It's interesting that it does take time even when there's something that when I look back I say, oh, it must have been an eye opener for everybody right away. The math was difficult, and it took time for people to learn it. Don't forget that in that era, optics, electric, and magnetic studies were three distinct branches of physics. You learn optics, and that's your specialty. Somebody else learns electricity. That's their specialty, and these people have to learn to talk to one another and speak the same mathematical language before they can realize that there is an awful lot that this other field can now help you with.

The experimental verification happened about 20 years after the publication, and it was Heinrich Hertz, a German experimentalist, a very young fellow, who had looked at Maxwell's publications and realized, we should be able to build an apparatus. It's not really all that hard to directly test this hypothesis. It's one thing to say it's obvious that light is an electromagnetic wave, but how do you know? How do you prove that there's a little electric field in the light? It all makes sense. It all fits together, but it's not yet direct observational, experimental proof. Heinrich Hertz did a much simpler experiment. He built a circuit, and in this circuit, he had a few components that would make electric charges slosh back and forth. It's called an oscillator. Charges go one way, and then they go the other way. You have a little power supply. You can drive the current. You can choose the frequency at which it oscillates back and forth, and now, according to Maxwell, any time you have oscillating charges, you should be creating this electromagnetic wave radiating outwards at the speed of light, and it's not necessarily visible light. This is very low frequency in Heinrich Hertz's laboratory so your eyeball wouldn't respond. He built another electric circuit, which was very, very similar to the first one. The new electric circuit is also an oscillator, but this one has no power supply whatsoever. It's just the oscillator with no source to make charges move around. According to Maxwell, this electromagnetic wave should spread across the room at the speed of light, and when it reaches the second one, there is this oscillating electric field. It points up. It points down, up, down. It's going to drive the electrons, and as soon as you have electrons in an oscillator, it's easy to tell because moving electrons heat wire up. They make light bulbs glow. It would be very easy to detect the motion of electrons inside of an electric circuit. That's what Heinrich Hertz did, and it worked like a champ, exactly as Maxwell had predicted.

Think about what Hertz has just done. This is an application of this crazy, abstract mathematical formalism, and he's just built a radio. You have a radio here because on the one side of the room, you have an antenna that broadcasts signal, and on the other side of the room, you have a receiver that receives the signal. If you just modulate the signal on one side, it's going to be modulated on the other. Maxwell's Equations tell us that if you jiggle the first charge twice as rapidly, then the other charge will jiggle twice as rapidly.

You're picturing what's going on here. A moving charge at one point in space creates this ripple of electric and magnetic fields, which allows other charges somewhere else to move. You push it. You could think of it as action at a distance again—a novel, a new kind of action at a distance, where motion of one thing makes motion of another thing, rather than just this simple push or pull like gravity. That's really just not the way Maxwell is thinking about it. He's thinking about it in terms of the traveling electric and magnetic fields in between.

This revolution in thinking about the world, this revolutionary unification and synthesis, took at least 20 years. When Heinrich Hertz did his experiment, I think that was for most physicists pretty much the clincher, and at that point people felt compelled to make sense of the mathematics, to learn it, and to recognize this revolution just as Copernicus did. Remember, Copernicus had published his heliocentric theory, and it was probably very obvious to him. It was a flash of insight. Of course, the sun is at the center, and the planets go around it. Then everything makes sense, all of the observations, all of the astronomy. It all makes sense if you just make this wild, revolutionary postulate, and yet, at the time, it was only a seed for other scientists such as Galileo, Kepler, and ultimately, Isaac Newton to recognize in retrospect as the moment of revolution. It was really the same story with Maxwell. The amount of time was much less. Instead of 100 years, it was only 20 years.

This kind of unification is, in a certain sense, every theoretical physicist's dream. It's your goal as a physicist to try to take disparate phenomenon, bring them together and realize it's all just one basic set of ideas. Newton did this, and did it in perhaps the most profound way. He said that what is happening here on Earth with marbles rolling down ramps in Galileo's experiment is really the same fundamental physics as the Earth going around the sun or the moon going around the Earth. It's all the same. It's all F=ma.

Now Maxwell has done this again. Everything that is electrical, everything that is magnetic, everything that's optical—it's all really ultimately just these Maxwell's Equations. He has not only done this wonderful service to the field of physics, but has also opened up new branches of physics, because now you can think about what might be called physical optics. You can think about optics in a new way. Instead of thinking about the path of rays of light through a prism, or how a telescope might focus the light, you can also think about the interaction of light with matter. It's just electric and magnetic fields interacting with charges in the matter. It's really all explainable by Maxwell's Equations.

Light had been studied forever. The Greeks were thinking about light. Isaac Newton was thinking about light. Many, many physicists had been curious about light. I would argue that Newton's career as a physicist began with his experiments on prisms and the colors of light. Newton recognized that white light was really composed of all of the colors in the rainbow, and Newton had a background working theory of light, which was that light was corpuscular. It was made of little light particles. If you have a light bulb, Newton was thinking that the light bulb is spewing out a bunch of little light particles in all directions. They travel outwards at the speed of light, and so Newton would think that when they hit your eye, it's some sort of F=ma thing where they're bumping into the retina in the back of your eyeball. He just has this different mechanism to think about light.

It turns out it didn't really matter so much because he wasn't paying much attention to the wave-like nature of particle-like nature. It was just his belief, but the experiments that he was doing had to do with other aspects of light, like the color of it. At the time, the technology didn't really exist for Isaac Newton to convincingly prove to anybody, including himself, whether or not light was a bunch of little particles, or light was some sort of undulation, some oscillating phenomenon. There were lots of theories. People were arguing with Newton back in his era, and it took until 1800, when a physicist named Thomas Young finally did an experiment directly with light to prove convincingly to the world that light is a wave phenomenon. We'll talk about that in a later lecture because Young's experiment is a little bit complicated, and it's really part of a broader story that we need a little bit more time to go into.

Young showed that light was a wave. This is 1801. Maxwell is not going to do his theory for another 60 years, so for 60 years the whole physics community is wondering what the heck is waving? Light is a waving, but what is waving? When we think of waves, we think of water. We think of some material object moving up and down, and nobody had any clue what could be waving. Maxwell answers the question. It's a new thing. It's the electric field that is waving. Electric and magnetic fields are moving. They're changing in strength and in direction. There's no material object moving at all. The wave is a wave of the electric and magnetic fields themselves. It's a wild idea, a little bit abstract. Again, you can recognize why it would be difficult for people to just accept this instantly, but it turns out to be enormously practical.

Think about lenses and how they focus light, or think about rainbows. Think about the colors that we see from the light, or think about a prism. Think about the defraction of light as it goes through a very narrow hole and spreads out. All of these can be described in the Newtonian era. They can be described, but they can't be fundamentally explained until we come to Maxwell and Maxwell's Equations.

Maxwell tells us that light carries energy with it. It's part of the equations. When you wiggle the electric charge here, you are doing work on it. You're applying a force. You're accelerating it. It's moving over some distance. Force times distance is work, so you're doing work on the charge. Whenever you do work, that means you're transferring energy. Where is the energy going? Conservation of energy—it's spreading out in this electromagnetic wave. Electromagnetic waves can carry energy. Well, we know that. Light can carry energy. Light from the sunshine carries solar energy with it, and so all of the sudden there's a whole new bunch of technology that's available to us—electric light sources and electro photovoltaic panels—although there is a little bit more quantum physics involved there. Fundamentally, the source of energy is pure, classical Maxwellian physics.

You're thinking about Maxwell's Equations, and remember the speed that he came up with, 186,000 miles per second; the speed of light, turns out is a number that is independent of any details of the motion. Maxwell thought really hard about this. What if you jiggle faster? What if you jiggle in a circle? What if you jiggle faster and then slower in a funny pattern? You will produce different ripples in the electric and magnetic fields. They themselves will wiggle at the frequency that you wiggle the electric charge. If you wiggle twice as fast, the electric field will respond twice as fast and the receiver, if you have one, the electric charge far away, will wiggle matching the original one. The speed at which this traveling wave spreads is always 186,000 miles per second, the speed of light. That's a fundamental property of light. What is the difference between different kinds of light? What is the difference between blue light and red light? All of the sudden Maxwell says, oh, I have it! Red light and blue light don't differ in their fundamental properties. They're basically the same exact thing; it's just that blue light arises when your retina wiggles faster. Red light is what you perceive when your retina has electrons in it that wiggle a little bit more slowly. There is this range of wiggle frequencies where your human brain can detect this electromagnetic wave.

What if you wiggle a little bit faster than what makes a blue signal to your brain? There is still an electromagnetic wave. It's still traveling in exactly the same way. It's the same physical thing, but you won't perceive it any more. It's beyond the violet. It's ultraviolet, and you would need some new detector that could detect this ultraviolet radiation. Sometimes I call ultraviolet radiation ultraviolet light because even though I can't see it, I think of it as the same thing as light. It's going to go through lenses. It's going to do all the sorts of things that light does.

If you go even faster still and wiggle your original charge faster than the ultraviolet, you move up into other kinds of electromagnetic radiation such as X-rays. X-radiation is just electrons wiggling even faster still producing a wave that's wiggling even faster still. Beyond the X-rays are gamma rays. X-rays are what the dentists use to image your teeth. The idea is, it's just light coming out of the X-ray machine. It's X-ray light. It's a different frequency of light, and when that light strikes your body, it turns out that it's wiggling fast enough that it can travel through much of your body. Most of the watery parts of your body are as transparent as glass to this particular frequency, this particular color of electromagnetic radiation. Your bones are dense enough that they will absorb this light, and so the picture that you see is a shadow picture. It's just like any shadow picture of light except that you need a different detector. Your eyeball can't detect the X-radiation. What if you go in the other direction? What if you wiggle slower than the red? You have the infrared. Infrared radiation arises from warm objects because warm objects have wiggling electrons in them, and they produce this electromagnetic radiation. It, again, is another form of light, and we can build cameras, night vision cameras, that detect that radiation. You simply need some electric circuit that responds in that frequency range.

In honor of Heinrich Hertz, we've named the unit of frequency the hertz. One hertz means one wiggle per second. The electric wall plug in your house is 60 hertz. It means that the electricity is sloshing back and forth 60 times a second. The light that your eyeball is sensitive to, visible light, would be about a million billion hertz. That's how many times per second electrons have to wiggle in order for your biological retinal cells to produce an electrical signal to your brain.

If you go beyond the infrared, you are into the microwave radiation range. These are just names that people have given to different frequency bands. Microwaves in your oven are nothing more than another kind of light. It's just electromagnetic radiation at a much lower frequency. It happens to be the frequency at which water molecules like to jiggle, and so this radiation tends to make water molecules wiggle back and forth. That heats up your food.

If you go even lower in frequency still, you are down to what we call radio waves. The radio waves that are being broadcast are just a big antenna somewhere with an electron that goes up and down that antenna, an electromagnetic wave that travels out, reaches your antenna, and the electrons at your house go up and down. It's just Heinrich Hertz' experiment all over again. The only reason you need to plug in your radio is to amplify the signal. You take your speakers and make that oscillating signal push the speakers in and out at some frequency. You need to change the frequency so that it's even lower still so that it produces a wave that your ears can detect. We'll talk about that in the next lecture.

In the end, you can see that Maxwell's Equations are not only leading us to new physics. They're also leading us to many, many new applications. Maxwell has helped create the field of applied physics and he has also opened up a whole new way to study optics and light. Think about television sets, cell phones, garage door openers. They're all Maxwell's Equations, and what I want to do now is to think more about waves in general. Now we see how important waves are in the electromagnetic world. We'll realize that waves are important everywhere.

Lecture Nineteen
Vibrations and Waves

The wireless telegraph is not difficult to understand. The ordinary telegraph is like a very long cat. You pull the tail in New York, and it meows in Los Angeles. The wireless is the same, only without the cat.
—Albert Einstein

Scope:

In this lecture, we step back from the story of electromagnetism to think about a very different kind of physics—the description and understanding of objects that vibrate and the associated phenomenon of waves. Vibrations and waves are *everywhere* in the natural world, and they provide a wonderful counterpoint to our usual language and model of particles. Understanding the big ideas of waves, especially the remarkable feature of *interference* and the point-counterpoint of waves versus particles, plays a key role in the developing story of physics.

Outline

I. Vibrations and waves are everywhere in the physical world and provide a counterpoint to our usual language and model of particles.

A. Waves are a collective phenomenon, a way of seeing a simple pattern in complex situations. They have been studied since before Newton and were well known by Maxwell's time.

B. The "canonical" wave would be a pebble thrown into a pond, resulting in ripples of water spreading out. How can we *describe* this?

1. The water itself is the medium for the wave. The water molecules are being displaced from their equilibrium point; as the wave passes, they move up and down. In other words, the ripples that we see on the pond arise from the displacement of particles from their normal equilibrium level.
2. A wave is not a "thing" itself; it is a self-propagating disturbance. A classical wave has no obvious physical essence—no mass or clear position.

3. If you look at the ocean, you see waves spread out parallel to the beach, one after another. This motion is not localized at all; the entire ocean carries gigantic, spread-out traveling waves.

II. A subtle phenomenon takes place in the motion of a wave.

A. Picture a wave traveling from left to right. The water—the medium of the wave—is moving *only* up and down; it is not traveling sideways.

B. Water waves near the beach, where we most often see them, become nonlinear—the water splashes sideways—and the waves are no longer ideal classical waves. If you were sitting in a dinghy beyond the break, however, you would see that the waves cause you to bob up and down on the water; they do not cause you to start surfing.

C. Think of a field of wheat, disturbed by a rustling at one end. The wave spreads out and travels across the field, but no wheat stalk ever leaves its original spot. The wave travels across the field, but wheat does not!

D. Consider the "wave" in a stadium, when people rise up and down in their seats in a sort of contagious motion. The wave rushes around the stadium, but the people (the medium) ultimately stay in their seats.

III. Waves may seem to behave in some respects like particles, but the two are not the same.

A. Particles have mass and position and exist independent of any other material objects. None of these is a characteristic of waves.

B. Let's think about a Slinky® to visualize some of the differences between waves and particles.

1. A Slinky stretched out motionless on the floor is the medium.
2. Imagining holding one end of the Slinky fixed and jerking the other end up and down one time. A sideways pulse will travel from one end of the Slinky to the other.
3. The pulse almost seems like a material object; it obviously travels from one end of the Slinky to the other and will even recoil and travel back. With this behavior,

it's easy to think that the pulse is somehow like a particle.

C. Waves are characterized by a frequency (measured in Hz), a wavelength (the distance from one peak of a wave to the next), and a velocity (the speed of the wave itself, not the medium).

1. The speed of the wave arises from interactions of the medium. Generally speaking, the more tightly coupled the "pieces" of the medium are, the faster the wave will ripple.
2. If you jerk one end of a Slinky quickly (rapid frequency), the pulse will have a different shape (wavelength), but the pulse's progress as a traveling wave will not be affected. (See the Essential Computer Sim at the end of this lecture.)
3. If you stand up and sit down quickly in the stadium, you will affect how wide the stadium wave appears, but the traveling speed of the wave has to do with the interaction of you and the person in the seat next to you, not with your behavior alone.
4. Maxwell saw this clearly with electromagnetic waves, which always travel at the speed of light.

IV. Waves are closely related to *simple harmonic motion* (*SHM*)—oscillations.

A. The mathematics of SHM is described by the *sine wave*, which represents something moving back and forth smoothly, forever.

B. SHM is ideal oscillatory motion. Do real objects in the world behave in this ideal way? To a large degree, the answer is yes. Electromagnetic waves are truly ideal; the Slinky and water waves away from the beach are fairly close approximations of SHM.

C. What makes such motion, and why is it so common?

1. Any material object that has a "home," an equilibrium point, is pulled back to that point whenever it is displaced. The result is generally SHM.
2. Think of a guitar string that you pull away from equilibrium, creating tension in the string. As the string is pulled backed toward its original position, the

principle of inertia takes over (an object in motion remains in motion), and the string moves past its equilibrium point. Then, of course, it's pulled back toward its original position again and so on.

3. SHM takes place throughout the universe, for example, in the Earth orbiting the Sun (viewed from the side) or atoms moving in a crystal.

D. How do we know when a wave is happening?

1. If we zoom in on a wave, we see the SHM of the medium.
2. If we zoom out, we see a wave traveling at some speed, and we no longer pay attention to the medium. We don't see the SHM; in fact, any point on the crest of the wave seems to move in a straight line, like a particle.

V. A defining characteristic of a wave is what happens when two waves come together.

A. Picture the Slinky again, with a person holding each end. If each person creates a pulse and two waves begin to travel along the Slinky from opposite directions, what happens when they meet?

B. If the waves were particles, they might break or bounce off each other, but in fact, the waves pass right through each other.

C. The most interesting phenomenon takes place at the point of intersection of the two waves; this is called *superposition* or *interference*.

1. If I send a 1cm tall "up pulse" along a Slinky and you send an "up pulse" from the other direction that is also 1 cm tall, at the point where they meet, we momentarily have a pulse that is 2 cm tall.
2. If I send an "up pulse" and you send a "down pulse" from opposite ends of the Slinky, at the point where they meet, we momentarily get complete cancellation.
3. *Constructive interference* takes place when two pulses add up, *destructive interference* is when they cancel each other out.
4. Think about how dramatic destructive interference is: We would never have two particles coming together,

briefly disappearing from the universe when they meet, then reappearing after their interaction.

VI. No matter where we look in nature, we see oscillatory motion—the behavior of waves. Next we'll talk about some more specific examples of waves.

Essential Computer Sim:

Go to http://phet.colorado.edu and play with Wave on a String. You can use this sim to help answer several of the questions below. With this sim, you can also explore reflections, pulses, and the relationship between wavelength and frequency and learn about what affects wave speed. Also try Masses and Springs to learn about simple harmonic motion. Sound is a good simulation to get a sense of the propagation of waves and the geometry of interference in two dimensions. Going a little further afield, you can investigate the mathematics of sine waves with Fourier: Making Waves.

Essential Reading:

Hewitt, chapter 18.

Thinkwell, "10: Oscillatory Motion" (first segment on simple harmonic motion) and "11: Waves: The Basics of Waves."

Recommended Reading:

Crease, chapter 4 and start of chapter 6.

Questions to Consider:

1. If you stretch a Slinky and wiggle one end, a pulse or wave will travel along the Slinky. What must you do to change the traveling speed of this wave? What is it about the wave that changes if you wiggle your hand faster? (The answer might surprise you. Give it a try. If you can't find a Slinky, go to the Wave on a String simulation and see if that helps you answer the question.)
2. Can you devise a real or simulated experiment to convince yourself that two pulses traveling oppositely on a Slinky combine to make a doubly big pulse? What happens if the two pulses are not the exact same shape?

3. How many examples of oscillations can you think of in everyday life? Are they all pure SHM, or are some oscillations more complicated? (Is the motion of a piston in your car engine SHM?)
4. Can two traveling waves, moving in opposite directions, reflect off of each other? Why or why not?
5. If two waves head toward one another with opposite "signs," they destructively interfere at the point of intersection. Does this mean that some energy was temporarily destroyed, only to reappear later (when the waves continue along)? If not, where did the energy go at that special point?

Lecture Nineteen—Transcript
Vibrations and Waves

I'd like to step back briefly from the story of electromagnetism to think about a broader kind of physics, the description and the understanding of objects that vibrate. We could think of a pendulum. We could think of a guitar string vibrating back and forth—anything that vibrates—and the associated phenomenon of waves. Vibrations and waves are everywhere you look. Just think about it. It's happening at a microscopic level. It's happening at a macroscopic level. Many, many objects in the world wiggle back and forth, and when we start to think about how waves work and how oscillations work, we'll discover that we have a new language for thinking about things and events in the world that's a nice counterpoint to the language that we've been stuck with, which is the language of particles. We've been so focused on particles because that's where Isaac Newton started us. We were thinking about the world is made of objects, and that's a good, constructive, productive way to think about things.

The study of electricity and magnetism has maybe led us to think about nature as having this field aspect, and fields are more continuous. Then, you have these wiggles in the fields and suddenly, you realize that understanding waves might be useful to be able to describe what's going on in a broader sense. It's really a collective thing. If you look at one object, well, there you have your point-like object. It's when you have a collection of objects that you start to think about the wave. That's where waves arise. They always arise when you have a bunch of connected things, and then you have some disturbance that can propagate and spread. People have been studying waves forever. Certainly, ancient Greeks were interested. It's natural human curiosity to look at the waves on the beach and to wonder about them, and, boy, it looks like physics. It seems as if you should be able to describe this thing. You should be able to understand how it works. They seem to be repeatable and regular, and you can draw pictures of them. It just feels as if this should be good physics.

By Maxwell's time, waves were extremely well studied. It was part and parcel of what it meant to be a physicist. You knew about Newton's laws. You knew about waves, how they behaved, what they looked like and how you made them. They were definitely physical things. You could see them with your eyes. You could measure their properties, and so we ought to be able to make sense of the story. I will always go back to my canonical wave, which is my dropping the pebble into the pond. I like that one. It's easy to visualize. You can see in your mind this little ripple spreading out, and you can start to think about the ways that we might use to describe this. Physics begins always with description. Kinematics was the beginning of classical physics so I'd like to discuss the kinematics of waves. What do we need to just describe this thing? Then, once we can describe it, we'll start moving to deeper understanding and explanations.

If you look at the water and you think carefully about what's happening, there is a medium. There is this material. That's the water itself, and the water molecules are moving up and down. They are being displaced, and they have an equilibrium point. They want to be at the flat level of the pond. If you push the water down, water pressure lifts them back up again. If you pull them up, gravity pulls them down.

The ripples that you see are arising from the displacement of the particles from their ordinary position, and that's really all we're going to need to describe, but when we step back and we see the wave, we don't have to think about the water. You can squint your eyes a little bit and not worry about what the water is made of and just watch the wave. How big is it? Where is it located? How wide is it? These would be the kinds of things that we would want to do to describe it.

In the end, I will define a wave as a self-propagating disturbance of anything. It's self-propagating—it has to create the next piece of itself. So the water molecule here wiggles its neighbor, which wiggles its neighbor, and that's the self-propagating disturbance. The wave isn't a thing. It's not a piece of water or a bunch of pieces of water. The wave is the disturbance of the water. It's one step removed, and we're going to wrestle with this a little bit. Is wave a thing or not? When I have a classical wave, an ideal wave, I almost always think about the medium, although we already have this one example of an electromagnetic wave where there is no physical thing, no material object, that's wiggling. It's the electric field, which is an abstract entity, which is just getting stronger and weaker over and over again.

When I say it's not a thing, what would the mass of a wave be? It doesn't make any sense. The wave doesn't have mass. The water molecules have mass, but the wave, the disturbance, is just what it is. Does it have a position? Well, that one is a little bit touchier. If you think about my water wave that is expanding, I'm seeing this little ripple that heads outward. It does have a distance from the origin. At any given moment in time, I could take a snapshot and I could say here is the wave, and I would point to a big circle. I would know what the radius was, but I wouldn't know exactly where the wave is. The waves are always spread out. It's a feature of waves that they tend to be spread out. If you go out to the ocean, you can see a wave, and as they come toward the beach, they're spread out parallel to the beach. Then behind them, there's another wave, and behind that there's another wave. If you take the grand picture of the ocean, you'll see this undulation traveling across the surface of the water. It's not localized at all. The entire ocean is one gigantic traveling wave.

The location of a wave may or may not be well defined. It depends on the situation, and if you think about the motion of the wave, there's a really important and subtle thing going on. Picture a water wave traveling by, and it's moving from left to right. You're out in the ocean, and you see this big wave. It's very clearly identified as it travels along. Now, think about the medium, the water. The water is going up and down. Now, you might think that the water is also moving from left to right. If you've surfed at the beach, you might imagine that the water is flowing sideways right along with the wave, but it's not. I want to convince you of this because it's an important idea about the waves. Water waves near the beach unfortunately become non-linear. They are no longer ideal waves, and there is some throwing around of water at the beach, which really mucks up with our intuitions because we think of waves, most of us, by our experiences at the beach. If you just go a little bit away from the break, if you go to the region where they're just big waves but they're not breaking and you're sitting in a little rubber dingy, when the wave goes by, you go up and then you go back down again. You don't start surfing. You just bob, and if you're not convinced of this, I have another example that convinced me. I remember that I used to struggle with this one a lot.

Think of a field of wheat, and it's completely smooth and flat. Then over at one edge of the field, there's a rustle—maybe a little puff of wind or a person walks by and shakes the wheat. Now you're standing up above on a platform and you're watching. You can just visualize this wave of wheat. It's just a ripple in the wheat that's going to spread from one side to the other. It's very clear, very visible, and you could take pictures of it. You could say, ah, there it is now, and now it's moved over here. You could see the wave as it moves from one side of the field to the other. Now I ask you—does the wheat move from one side of the field to the other? Is there a flow of wheat just because of this wave? Now, it's completely obvious that every stalk is stuck right where it is. It just wiggles back and forth and ends up right exactly where it started.

Water is really doing the same thing out in the deep ocean. Most waves are ideal in this sense that the motion of the medium is one direction and the motion of the wave is in a totally different direction.

Here's another example, and one that many people have personal experience with—the wave at the stadium. Somebody stands up and whole rows of people all stand up and then sit down. Then their neighbors stand up and sit down. Then their neighbors—it's a wave. It's a self-propagating disturbance. You wait until your neighbor moves. There's an interaction. It's a social interaction. It's an agreement between you and your neighbor that now you will stand up, and people are going up and down. No human being is moving in the direction of the wave. The wave is rippling sideways through the stadium, but nobody makes one micron of movement in that direction. That's a property of waves that the medium is doing one thing and the disturbance is doing another.

I want to recognize again that the wave is an abstract thing. It's our description of what is going on. It's not a physical entity. It's not a particle. Now, if you watch a wave and look at it, you might think that it's behaving in some respects like a particle. Let's think about the differences between these things. A particle has a mass. It has a position. It has an existence independent of any other material object. None of these are characteristics of waves. Waves require a bunch of objects that are connected together, and the waves are spread out and can have all sorts of different ways of being spread out.

It is nice to think about a wave that is localized. This is going to help me visualize what's going on. My favorite example for this one is a Slinky. Remember those toys—I had lots of Slinkies when I was a kid. If you stretch out a big, long Slinky on a flat floor, at first it's motionless. That's the medium. The medium is the metal of the Slinky. Now, you go to one end. The other end is held fixed, and at your end, you start to wiggle the Slinky. First of all, just wiggle it up and back and stop wiggling. Can you visualize what's going to happen? There will be this little blip on the Slinky, a sideways blip, and that blip will start propagating because, of course, the sideways pulled metal is going to pull on its neighbor, which pulls on its neighbor, and of course, those neighbors are pulling the blip back. You have a restoring force and also this propagation. This is a beautiful little wave. It's a little pulse, and it travels from one side of the Slinky to the other. If you squint your eyes, you might say, well, that looks like a little toy something, a little toy truck that's moving from one side to the other. If I weren't looking too carefully, I might fool myself into thinking that this little blip was a material, physical object. It does things that material objects do. It goes to the end of the Slinky. It will recoil and come on back. It does other things that aren't common for material objects, such as it will fade away over time because of internal friction.

You can see that there are both connections and distinctions, but you can fool yourself into thinking that a wave is somehow particle like. We'll keep coming back to this because in nature many wave phenomenon people have argued—is that a wave, or is that a bunch of little particles as when Isaac Newton tried to decide whether light was a wave or a bunch of little light particles flowing outwards?

When you're looking at a wave and you want to describe it, one of the things that you might describe is how to go down to the level of the medium itself and see the medium wiggling up and down, up and down. And you say, okay, that has a frequency. I can measure that in hertz, cycles, or wiggles per second. The frequency would be a description. Is that a description of the medium or the wave? Well, they're both intimately connected with one another. When we're talking about a particular wave, it will have a definite frequency associated with it. It will also have usually a characteristic length. If you think about those waves on the ocean, as long as you have one, and then another, and then another, there's going to be a typical distance from one peak to the next peak. We call that the length of the wave. Be careful—it's not the sideways length of the wave running toward the beach. It's the distance from one peak back to the next one. That's the wavelength, and then you can talk about the velocity of the wave.

Now we're going to be very careful. We're talking about the wave, not the medium. If you go to the stadium and you watch the ripple, it's the speed at which the ripple moves sideways through the stadium. That's what I mean when I talk about the speed of the wave. I do not mean the speed of the people who stand up and sit down again. That is more closely associated with the frequency of the wave. We keep these ideas separate, and the speed is the thing that really belongs to the wave itself. The speed arises because of interactions. The more tightly coupled the pieces of the medium are, generally speaking the faster the wave is going to ripple because when you jerk on one piece, if it's tightly coupled to its neighbor, the neighbor is going to respond very quickly. That's how you make a fast wave is that you make things that are more tightly connected.

If you take that Slinky and you jiggle it and get a wave that's traveling down it and you want to make that wave travel faster, what do you have to do? The one thing that you might imagine is just jiggle your hand faster, just jerk it forwards and backwards in half the time. That won't do it. What you'll get is a wave that has the medium jerking up and down faster, but the pulse will continue to travel at the exact same speed as it did before. You may believe that. You may not believe that. You have to go try it. Buy a Slinky and check it out. I have a computer simulation of this. It's on a website from Colorado, phet.Colorado.edu, and you can look at lots of simulations of physical events. One of them is a wave on a string, and you can convince yourself at least with the simulation that jiggling your hand more rapidly will change the frequency. It might change the wavelength because if you jiggle up and down really quickly and then up and down really quickly over and over again, then there's not time for the first pulse to run away from you before the second pulse starts running away from you. The two pulses are going to be close together. You're going to have a small wavelength if you have a rapid frequency, but it has no impact on the speed at which those ripples are traveling away from you. It's a little bit counter intuitive, but a fact of nature that you can verify for yourself if you want.

If you stand up and sit down at the stadium very, very quickly, it won't help make the wave go faster. What would make the wave go faster is if your neighbor's reaction time was shorter. Your neighbor has to react to you more quickly. They have to be more tightly coupled. If you want the Slinky wave to go faster, what you do is you tighten the Slinky. You pull. You stretch it out. Now all of the pieces of metal are under more tension, and when you jerk one piece, now those little pulses will travel more quickly.

We also saw this with light. Mr. Maxwell proved that no matter how you jiggle the electric charge, the traveling electromagnetic wave always travels at the speed of light. In this case, there are no material objects involved so there is no mechanism to speed up the speed of light. We're stuck. It is what it is, and you can't make it go any faster.

Waves are very, very closely related to this oscillatory motion of the medium. The one is really part of the other, and physicists love to look at this kind of simple oscillation. We call it simple, harmonic motion. Simple because it's just something going back and forth, back and forth, and harmonic because that makes us think about music and musical strings—something that just moves in a nice, simple, back and forth pattern.

The mathematics of simple, harmonic motion is described by a mathematical function called the sign function or the sign wave. If you were to make a graph of the sideways position of a spot of water, or a spot of Slinky, or one of the people in the stadium, I really need a continuous wave. I need to keep wiggling the end of the Slinky. I need to have a whole bunch of waves going by. In the stadium, we don't usually do that, but you could imagine somebody rises up and sits down. Then they rise up. Then they sit down. Now you'd have more like a water wave, a continuous wave. Now, if you plotted the position of any one individual piece of the medium and you charted it on a graph, you'd see it goes up and down, and up and down in a nice, smooth sign wave. Simple harmonic motion is the name that we give for this ideal oscillatory motion, and you might ask—well, are real things behaving in this ideal way? The answer is it's a darn good approximation in a huge number of circumstances. It's remarkable how many things have motion that is practically indistinguishable from simple, harmonic motion. The Slinky is a pretty darn good simple harmonic motion for the individual pieces. The water waves work really well until you go right next to the beach. In electromagnetic waves, it's truly ideal. Electromagnetic waves are the perfect wave, and the motion, if you graph the strength of the electric field at some point, would be this beautiful sinusoidal wave.

What makes such motion? Why is it that it's so common? Any time any material object has a place where it belongs, where it has an equilibrium point and you disturb it and it is pulled back there—no matter which way you disturb it, it is pulled back home, and that's going to make simple harmonic motion. It's the guitar string that you pull away from equilibrium. There's now tension in the string that's trying to restore every piece of metal back to that flat position. Of course, they've been pulled all the way and now they're moving. Inertia keeps them going. An object in motion remains in motion. That's true for little pieces of metal. They're all connected one to the other, but each one of them is obeying Newton's laws. They keep on going, and they go past the equilibrium position, but as soon as they go past, they're being pulled backward. They are pulled forward, and then they are pulled backward. Then they are pulled forward. Then they are pulled backward, and it goes over and over again. If there's no damping, if there's no friction in the rubbing of these metal molecules one with the other, metal atoms, then this will keep on going for an arbitrarily long period of time. This oscillation can keep on going. Ideal simple harmonic motion never stops. In real life, it tends to fade away in amplitude if there's friction involved.

A pendulum would do the same thing. You lift it up. It is pulled back down to the origin, and then it has some inertia and swings on through. You can start to think about all sorts of circumstances in life, atoms in any crystal microscopically. If you pull them aside, all of the rest of the crystal is sitting there, and so they are pushed back to that spot where they are as far away from everybody as they can be. Of course, they'll overshoot so they also will undergo simple harmonic motion.

You will have a wave any time you have simple harmonic motion and one object that's doing that is coupled in some way to another object that can do exactly the same thing. These two things in nature, simple harmonic motion, or oscillations, and waves just come together, and we see it all the time. If you move and your neighbor moves, that's all you need to make a wave start. You can think about all sorts of physical situations where this might happen, and in the next lecture we're going to talk about some situations where it happens where it's not so completely obvious. I want, in this lecture, to think carefully about how we would know for sure that we have a wave happening.

When you have a wave, you can look deep or you can step back. If you zoom in, what you see is the motion of the medium, and it is simple harmonic motion. If you step back, you see a traveling wave with some speed, and you no longer think about the medium. You no longer pay any attention to the medium. You just notice the disturbance itself, and all of the sudden it feels really different. It doesn't feel like simple harmonic motion. I'm watching this little blip at the top of one particular wave crest, and it seems to be moving in a straight line like a particle. You can see how this particle-like aspect is there in the collective motion.

What is it about a wave that's particularly unique? Well, I would argue that one of the most remarkable things, one of the most spectacular things about the fact that it isn't a particle, is what happens when two waves come together. Let's start by thinking about the Slinky, because that's a nice, clean example. You don't have to worry about things spreading out in multiple dimensions. A Slinky is a one-dimensional wave. It's just a pulse that travels down a nice, beautiful, straight line. Imagine this long, long Slinky, and you're at one end and you make a blip, a pulse. It starts traveling, let's say from right to left, and I'm at the other end and I make a blip and that makes a pulse that's going the other way. We have these two pulses, and they're heading toward one another. At first, they can't possibly know about one another. The blip over here and the blip over there have no mechanism yet to communicate, but as they get closer and closer together, you realize that at a certain point they're going to touch. There is going to be some kind of interaction, and what is that interaction?

Well, once again this is a question for experiment. You can imagine what might happen. You can imagine lots of things. If you're thinking like Newton, you're thinking that these are particles, and you might imagine that they will bounce one off of the other. The incoming pulse will turn around, and the other incoming pulse will turn around. That's not at all what happens so you have to watch carefully. What you'll see is they just pass right on through one another, and after they have passed by there is no sign of the interaction. They just slipped right through one another. That's spectacularly not particle-like. Particles coming together head on will crash. They'll either stop or they'll break. Or they'll go flying in some other direction, but whatever it is that particles do, they're definitely not just passing through one another.

You might be wondering how I know that they pass through one another. Maybe they bounced. I mean, after all, I see a blip coming in and I see a blip going out. How do I know which is which? Well, think about an experiment. You could convince yourself quickly. When you make your blip, make yours twice as big as mine so you have a doubly big pulse coming in one way. I have a singly big pulse coming in the other way, and they pass through one another. The little one just keeps on going, and the big one just keeps on going. It's clear as day so you can really convince yourself that this is the way waves behave.

Now comes the most interesting part of wave behavior. This is the part that is the most distinguishing characteristic of waves that is very, very different from particles. Even more than passing through one another is what happens at that moment when they pass by one another—not afterward, but at the moment of interaction? We have some fancy names for this. I've often been using the word superposition. When you have two separate motions or two separate things, like two forces or two fields, and they happen to both be at the same place at the same time, they add up. Waves superpose in this way also. If you send in a one centimeter tall up wave and I send in a one centimeter tall up wave, at the moment, at the place and time, where they pass through one another, if you take a snapshot what will you see? Well, you'll see a Slinky, and the Slinky has to be somewhere at every point. It has to be located somewhere, and you'll see the peak of the Slinky is two centimeters tall. The blips have added, and one plus one is two.

What would happen if I added an up pulse coming from one direction to a down pulse coming from the other direction? You can jerk the Slinky either way, and you don't have to make a full oscillation of your hand. You can just make a little jerk down and back. Start at the equilibrium, and that makes a little blip that just goes down on one side and just up on the other side. These two pulses are coming toward one another. After they pass through one another, we know what's going to happen. They're just going to keep on going. The down one keeps on going. The up one keeps on going, but what happens at that moment of intersection? If you take a snapshot, what happens when you add positive and negative? You have zero. It's quite impressive. It's as if the wave just disappeared for an instant in time. If you took a snapshot, you would see a completely flat Slinky. Now, that might seem paradoxical. Have we lost conservation of energy here? I mean the waves are carrying some energy and some information. Oh, it's still there. It's like taking the snapshot of a pendulum that just happens to be passing through the origin. If you took the picture, you'd say, oh, it appears to be at its equilibrium rest position, and it is at the equilibrium rest position, but the pendulum is still moving. The Slinky is really still moving. Parts of the Slinky are moving down. Parts of the Slinky are moving up. You just happened to catch them at this lovely moment where the two waves interfered with one another and apparently canceled one another out. Super-position is one name that we give to this adding up of waves. Interference is another name that we give. I don't know if I like the word interference as much, although everybody uses it, because interference makes me think that afterwards they will be different, that they interfered with one another. Interference makes me think of things being stopped, but it's just a word and what it means is super-position. You can have constructive interference, which is the two up pulses adding up to a bigger one or you can have destructive interference, which is where one is up, and one is down and they cancel out. Then, of course, you can have anything in between. You can have partial cancellation with wave.

The really dramatic thing, and the way you can tell a wave is a wave is that you could never conceivably imagine two particles coming together and then for one instant in time completely disappear from the universe and then reappear and continue on their way. It just doesn't happen. This is going to be our defining character of what makes a wave a wave, and when we're thinking about things like electricity and magnetism or sound waves or other kinds of waves, we're going to look for this destructive interference. That's going to be our signal that we have waves.

To summarize, we talked in an abstract way about waves and simple harmonic motion today. It's just a description of nature. It's a very, very common motion. Every atom in your body behaves this way.

When you drive the speaker, the speaker heads move back and forth, and then your eardrums move back and forth. Then when you look at me, your eyeballs have electrons in the retina that jiggle back and forth. No matter where you look and no matter what you think about in nature, you start noticing oscillatory motion. It's obviously important to describe it, and it turns out that physicists love this because the sign function is such a relatively easy mathematical description that applies to all of these different things. It's easy to manipulate them. It's easy to understand them, and we'll continue on this story next time as we talk about some more specific examples of waves.

Lecture Twenty

Sound Waves and Light Waves

Interference in ordinary language usually suggests opposition or hindrance, but in physics we often do not use language the way it was originally designed!

—Richard Feynman, from *The Feynman Lectures in Physics*, vol. I

Scope:

What do physicists mean when they say that sound is a wave or light is a wave? In this lecture, we will consider this question. Newton (of course!) was one of the founders of modern optics. Although he conceived of light as a stream of particles, his classic and beautiful experiments with prisms and optical lenses led to both theoretical and practical understanding of light that lasted for a century. More than 100 years after Newton, Young turned the world of optics on its head when he convincingly and dramatically demonstrated that light was *not* made of particles but was, in fact, a wave phenomenon. Maxwell's theoretical triumph 50 years later, showing light to be an electromagnetic wave, tied the story off neatly. Showing that Newton was wrong about *anything* is inevitably "revolutionary" but in a very different way than Newton's "revolution" from Greek natural philosophy; Young's revolution altered and deepened our conceptions of physical phenomena without breaking the structure of physics itself.

Outline

I. What do we mean when we say that sound is a wave?

A. Consider first a model for sound waves propagating in a medium, in this case, air.

1. Think of air as a collection of tiny, independent particles, like superballs, flying around the room and bumping into one another and the walls.

2. If you clap, you compress these superballs locally, creating a region of high pressure. The superballs will push against their neighbors, which in turn, extends the high-pressure region. The result is a traveling wave of compression, a disturbance of the pressure of the air itself.

B. The alternative to this wave model of sound might be one in which clapping results in the release of some kind of particles of sound.

II. What experiments could we do to determine which model of sound—the particle model or the wave model—is more accurate?

A. We might, for example, try to measure the speed of sound, but doing so would not allow us to conclude whether we were dealing with particles or waves.

B. At an outdoor concert, you might notice that even though you are some distance from the musicians, you still hear low frequencies and high frequencies at the same time.

1. This suggests that sound is a wave because, as we learned in the last lecture, both high- and low-frequency waves travel at the same speed.
2. If sound had a particle nature, we might think that high-frequency sounds would correspond to higher energies, which would mean that these particles would travel at higher speeds.
3. This argument is compelling, but not completely convincing, for determining which model is correct.

C. Waves should be "wavy"; that is, if sound is a wave, it should, for instance, bend around corners, and in fact, it does. You can easily hear someone in another room even if the room is around a corner from where you are. Still, sound might be particles bouncing out of the room, through the doorway, and around the corner.

D. To test our model, we might set up an experiment with a microphone, which is nothing more than a little flap of material that moves with the alternating high pressure and low pressure of sound waves. The microphone converts the motion of this little flap into a voltage.

1. We could watch the output of the microphone on an oscilloscope, and we would see the beautiful sine wave pattern that we talked about in Lecture Nineteen.

2. Is this proof that sound is a wave? Not completely, because sound particles might be striking the flap, which then "rings" like a bell.

E. As mentioned in the last lecture, the most distinctive feature of waves is the characteristic of interference. How could you observe this characteristic with sound waves?

1. You might set up two speakers, one in front of you on the left and one in front of you on the right. You stand at a point equally distant from both speakers while they both broadcast the same steady tone with the exact same loudness and pitch. At low frequencies, you can actually see the cone of the speaker moving in and out, but this is still not proof that we're dealing with waves.
2. You might then reverse the wires on one speaker so that when one speaker is pushing out, the other is pulling in; the speakers would be precisely out of synch, or *out of phase*.
3. In the wave model, when a speaker is pushing out, it's creating high pressure, and when it's pulling in, it's creating low pressure. Because the sound waves started out of phase and travel an equal distance to you, they will still be out of phase, and they will destructively interfere at all times. You should hear nothing.
4. In the particle model, the speakers spew out sound particles, whether they are in phase or not. The sound you hear, then, should be twice as loud.
5. Amazingly, if you performed this experiment (which is a little tricky), you would find that you could be standing in a room with two speakers broadcasting and hear silence. Audiophiles take this effect into account when setting up speakers in their homes.

F. We can also think of another experiment that might decide the question of whether sound has a wave nature or a particle nature: What if we remove the medium in which sound propagates (air)?

1. If sound is a pressure wave in air, then without air, there can be no wave. But if sound is made up of particles, the particles should still be able to travel through a vacuum.
2. This experiment was performed in the 1600s. Today, we could put a bell inside a jar and pump out all the air. As

the air is removed, we could see the clapper of the bell moving, but we wouldn't be able to hear any ringing.

G. No single experiment proves the hypothesis that sound is a wave, but we have seen that sound is not particles. The wave model seems to be consistent, and it enables us to make predictions about what will happen in our experiments.

III. It is more difficult to conceive of light as a wave.

A. This difficulty stems from the fact that the wavelength of light is very small, less than a micrometer. In contrast, the wavelength of sound is on a more human scale and is noticeable, as we saw when we talked about the room with two out-of-sync speakers.

B. Newton believed that light was made up of light particles, which is not a completely illogical hypothesis. Recall that sound can bend around a corner; you can hear a person in another room around a corner from where you are. But you can't see a person in another room unless that room is directly in front of you. This seems to be evidence that light is not a wave. (We now know that the smaller the wavelength, the less waves tend to bend around corners.)

C. In 1801, Thomas Young conducted a dramatic experiment in which he managed to see two light waves canceling each other out, much as we saw with sound waves in our speaker example.

1. Young used a bright light source, with the light passing through a partition into which he had cut a narrow slit. As the light passed through this slit, it spread out in all directions and illuminated the far wall uniformly. This setup is the equivalent of creating a point-like source of light.
2. Next, Young cut two slits in the far wall (this is the *double-slit experiment*). The expanding waves of light striking those two slits are symmetrical. Thus, the light at the two slits would be in phase (*coherent*).
3. Two expanding waves of light now travel through the back side of the double-slit partition, and there are now places further along in the room where they add up and other places where they cancel.

4. The result is a classic interference pattern—alternating bright and dark spots in a pattern exactly predictable from the wave model. Further, we can predict what would happen if we changed the color of the light, the number of slits, or the spacing of the slits.

D. Young's experiment clearly proved that light is a wave, because only waves exhibit this kind of "destructive interference." But a puzzle remained: What is "waving"? As we've discussed, Maxwell solved this puzzle 50 years later with the idea that light is an electromagnetic wave—that is, electromagnetic fields waving in strength.

E. As we've seen with other fundamental ideas of physics, once we know that light and sound are waves, we can begin to find practical applications for this knowledge, such as sound-canceling headphones and anti-reflective coatings on glass.

Essential Computer Sim:

Go to http://phet.colorado.edu and play with Sound (particularly the Two Source Interference and Varying Air Pressure tabs, both of which will help you with the questions below); also try Quantum Wave Interference.

Essential Reading:

Hewitt, chapters 19 and 28.

Hobson, start of chapter 8

Thinkwell, "11: Waves: Waves on Top of Waves," "11: Sound," and "11: Interference."

Recommended Reading:

Crease, chapters 4 and 6.

Questions to Consider:

1. If you know that energy is being transmitted from one place to another, what sort of experiments can you think of to determine whether the energy was being carried by particles (material bodies) or by waves?
2. Give two reasons why sound waves decrease in strength as they move away from the source. How would this compare to

electromagnetic waves leaving a source in empty space, and how would *that* compare to electromagnetic waves leaving a source in a region containing material (such as gases or glass)?

3. If you have a computer with a sound card, go to the Sound simulation at http://phet.colorado.edu and look at the tab for Two Source Interference. The listener starts at the exact midpoint. Do the two sound waves add up or cancel out? Move the listener around. Can you find the quiet spots where destructive interference is taking place? What can you say about the location of those spots? (How does the location depend on the frequency of sound?) Can you explain what is going on with the physical air (and air pressure) at those quiet spots?

4. Go to the Quantum Wave Interference simulation Start with photons, and turn the intensity and screen brightness up high. Choose Double Slits; make the slit width as small as possible and make the slit separation as wide as possible, with the vertical position in the middle somewhere (so that you're basically doing Young's experiment!) If you believed that light was a stream of particles, and you shined light from a pinpoint through *two* holes, what pattern of light would you expect to see on a far wall? Where would it be bright, and where would it get darker? Is there any reason to expect the light to be bright at a spot *directly behind the wall* at the center, that is, exactly between the two slits? Shouldn't that be a dark shadow? How does the wave nature of light explain the brightness right there at the midpoint?

Lecture Twenty—Transcript
Sound Waves and Light Waves

When I think of waves, I tend to think of the ocean. I visualize the waves on the ocean. Maybe because I'm a physicist, I might think about a slinky where you're jiggling one end and you can see these beautiful little waves traveling down the slinky. Today, I want to think about waves where it's much harder to see with the plain eye that what you have is a wave in front of you. I'm thinking specifically about sound waves and light waves, where you just have to argue scientifically that it's a wave without being able to see the medium moving. We're really going to be addressing the question, what do you mean when you say sound is a wave or light is a wave? It's an interesting question. We're going to be looking at some arguments that you might have between the wave point of view and the particle point of view.

Let's start off with sound waves, and let's think first about a model in which we have waves propagating in a medium. What would that medium be? Well, if I'm speaking to you, then the medium would have to be the air. I'm thinking now about the particles of air, the molecules of air, as little super balls. They're flying around, and they travel in straight lines. They could bounce off of each other. They could bounce off of the walls, but that's the way I'm going to visualize the room that I'm in. What would the wave be? Let's think about an example where we make a nice clear sound. I clap my hands, and what have I done? I'm now visualizing the air as filled with little super balls, little invisible super balls, and I've squeezed them. I've compressed them. I've created a region of high pressure momentarily, and now think about the consequences. If you have a bunch of super balls squeezed together, they are under high pressure. They are going to bounce against their neighbors, making their neighbors go into a little high-pressure region. Then they push on their neighbors, who are squeezed into a high-pressure region and so on. You push your neighbor, who pushes his or her neighbor, who pushes his or her neighbor, and that's exactly what you need for some sort of propagating wave. In this case, the wave is a pressure wave. It's an oscillation or disturbance of the pressure of the air itself, which is spreading outwards, and the individual air molecules bump into their neighbor and then bounce back again. It's not that when I clap my hands some air moves from me to you. That's not how it's working. It's this disturbance that is traveling from me to you. That is my wave model of sound.

The alternative, I suppose, would be a particle model in which, when I clap my hands, I spew out whatever it might be, little particles of sound, that go in every direction and ultimately reach your ear. We're trying to ask, how would we decide which of these two different models of the world—the wave model or the particle model for sound—is better?

When I think about this, I ask myself, what kinds of experiments could I do to decide? One thing I might think about is the speed of travel. You could measure the speed of sound. You have to think a little bit about how you would do this. You could go somewhere where there is a good, clear echo. You could clap your hands, and the sound travels out. It bounces off the wall. It comes back again, and you could measure the amount of time it takes for the sound to go out and back. You know distance and time, and that would determine the speed. But that's not really going to decide between a sound model and a wave model. In both of those schemes, which seem somehow very different from one another, you could still imagine that there would be a finite speed of sound. The little particles of sound have some speed or the waves travel with some speed.

If you go to an outdoor concert, you will notice that even though you're very far away, there might be a noticeable lag—you see the drummer hit the drum and then a fraction of a second later you hear the drum—so you're certainly far enough away that the speed of sound might matter. You will notice that the high sounds and the low sounds come at the same time to you. The music still sounds normal, so that indicates that high frequencies and low frequencies are coming toward you with the same speed. That makes you think maybe it's a wave because we've seen with waves that high-frequency jiggles travel outward and low-frequency jiggles travel outward at the same speed. For light, for electromagnetic waves, all frequencies travel at exactly the same speed, whereas for particles, if you were thinking about little sound particles, you might imagine that the high frequencies could perhaps correspond higher energy. Higher energy would probably mean higher speed or velocity, and so that seems like a possible inconsistency, but it's not a proof. You're hard-pressed to just based on this one argument decide whether sound is a wave or a bunch of particles.

The game that we're playing today is to try to fight it out between these two different models. Let's think about experiments that we could do. This is really the game of science. You have some thing that you're interested in, some question that you're puzzling over, and you create multiple hypotheses. Historically, there were many, many more than just these two. Once you have a wave model or a particle model, you can start to think about how to build that model in many different ways, and there might be completely different ways of imagining what sound is and how it works. What we're really doing is following consequences, thinking about experiments, and thinking about what would validate or invalidate the hypotheses. You might also wonder why I would care about this argument. Who cares whether sound is a wave or sound is a particle, and it goes beyond just curiosity, our search for the truth. We have this belief as classical physicists that there is a truth, a reality, out there that we are investigating, we're trying to make sense of, but it's also a practical thing. If sound is particles, I'm going to bottle them and sell them. You can open up the little bottle and hear an orchestra. That would be cool. If sound is a wave, then I could take advantage of the fact that I know, for instance, that waves can interfere and cancel one another, and I could make a lot of money by building a sound-canceling headphone. Well, somebody has done that, so you have a good clue about which is the correct or physically realistic model of sound.

I want to think through the argument because people didn't always know. If sound is a wave, it should be wavy. Think about how waves behave. If you look at water waves, when they start at a point, they tend to spread out in all directions, like dropping a pebble in the pond. If sound were a wave, you would expect that it should, for instance, bend around corners, go through small cracks, and then bend and go in all directions. It does precisely that. If you are in your house and the door is closed and you can barely hear the person in the next room over, and they open the door, all of a sudden you can hear them much more clearly. What's the physics? How do you understand that? Well, you can't necessarily see these people and the sounds that they're making because they might not be straight through the doorway. They might not be in a line of sight, but the waves go through the doorway, and the doorway acts like a small source of sound and the sound spread out in all directions. Sound can bend around corners. That seems like a fairly convincing proof that sound is a wave, but let me now defend the particle point of view.

Let me try to stick it out. Scientists do this all the time. You have your favorite theory and people are arguing against you, and you say maybe sound is still a bunch of particles spewing out but they're just bouncing all around. They're bouncing off the walls. They're bouncing off the door. They're bouncing off of the air, and so they make it to your ear on a more complicated path. That could explain how sound can go around corners and yet still be little sound particles that travel.

Now, we could pursue this. You could imagine making careful measurements of the amount of time it takes for a sharp sound to make it through a doorway. Let me leave that one, and let me think about other experiments that we might do.

We'll abandon that and say, all right, let me set up a little microphone. A microphone is nothing more than a little flap. It's a little flap of paper, or plastic, or material, and when sound waves hit it, the sound wave in my wave model is alternating high pressure, low pressure, high pressure, low pressure. That is the picture we have in this model, and if you look at the output of the microphone—microphones just convert the motion of the little flap into a voltage. You can look at the voltage on a screen, for instance, with an oscilloscope. It's a device that simply graphs voltage as a function of time, and you can see a beautiful sinusoidal wave pattern when a steady tone is reaching that microphone. You say, look, it's evidence. It's proof that sound is a wave because I see a wave on the oscilloscope screen, and that's pretty compelling until the particle believer says, well, okay I have an alternative idea. Maybe those little sound particles are hitting the flap and then the flap starts to wiggle. It's like striking a bell with a hard object. The bell will ring, and then if you're monitoring the bell, you might falsely conclude that the original source was wavelike.

This is an important point and it happens all of the time when you have scientific debates. You have to be careful, because what you are calling direct evidence of the thing that you're investigating, might be instead the apparatus that you're using to detect the thing that you're interested in. You have to be careful that you can separate the sound wave from the waving motion of the microphone. This is always true in physics. You have to be careful, and there again you could make some arguments and say, well look, if it's like a bell, bells have certain tones that they prefer. But the microphone doesn't have any preferred tones so it doesn't seem to be acting like a bell.

Let me leave these arguments and go to one more. Let's go straight for the convincing argument, because last time we said if you have waves, the real clincher—the thing that shows you that you have waves and not particles—is what we call interference, where two waves pass through one another. When two waves pass through one another, they interfere, and that word means that if the two waves are both up waves, then you have a big up result. If one wave is up and the other wave is down, they tend to cancel and you have no motion at that spot, in that time, at all.

How could we observe this affect? This would be a very clear evidence that sound is a wave if we see interference of two different sound waves. The simplest experiment that I can think of to do that would be to set up two speakers. We'll have one speaker in front and to the left and another speaker in front and to the right. Let's stand equidistant from both of them, and we'll drive both speakers with the same steady tone so they have the exact same loudness and the same pitch coming out of them. If you turn one of them off, you hear a nice loud sound. If you turn the other one off, you hear a nice loud sound. What happens when they're both going?

Let me play one little trick to make this especially easy. I'm going to take one of the wires going to one of the speakers and just reverse them, just flip the wires. Think about what happens. If you watch a speaker and you drive it at low frequency, you can see the cone of the speaker going in and out, in and out, in a lovely sinusoidal pattern. Now, is that proof that you're making a sinusoidal wave in the air? Not absolutely. Maybe the speaker cone is jiggling in and out whacking into the air and creating little sound particles that then fly out. Maybe they fly out in a high density and then they stop and then they fly out in a high density. You could imagine that there would be little wave-like observation from a wave-like source even though the thing that's traveling in between might not be a wave. I don't think that observation that the speaker is going in and out is proof that sound itself is a wave. The point of flipping the wires was that when one speaker is going out, the other is going to be going in.

The two speakers are going out of phase with one another or out of synch, and they're perfectly out of phase. When one is out, the other is in, and vice versa. Why would I want to do that? Well, if I believe in the sound is a wave model, then I'm going to argue that when the speaker is pushing out, it's making high pressure, and when it's pulling back it's making low pressure. I'm visualizing the sound wave coming from one speaker as alternating high pressure, low pressure, high pressure, low pressure coming toward me. It's a traveling wave front, a disturbance of the medium coming toward me. If I have a particle view, then it might be loud and soft, loud and soft, or it might just be a steady stream, but in any case, there are always particles coming toward me from both speakers. If I am a firm believer that sound is particles, I would argue that I really don't care whether the two speakers are in phase, out of phase, or completely random and disconnected form one another. If they're both equally loud, then I would expect to hear doubly loud sound at my ears if sound is a bunch of particles.

What if sound really is a wave? Then I have a wave coming from one speaker, and that means that the pressure at my ear is going high, and then low, and then high, and then low. Coming from the other speaker, it's going low, then high, low, and high, and I'm superposing these two waves. What happens when you superpose two waves that are exactly out of synch with one another? They cancel at all times. When one is high, the other is low. Plus one and minus one add to zero. Then, a half cycle later, instead of high plus low you have low plus high, but you're always canceling. You will hear silence. It's a remarkable thing. It's a very dramatic demonstration. There is a loud speaker over here and a loud speaker over there, and yet right where I'm standing it's quiet. If I move a little bit off to one side, then I'm no longer equidistant from the two, and if I move just a little bit, then one of the speakers will have gone through a little bit more of a cycle by the time that it reaches me and the two waves will be back in synch. Now I have high from one and high from the other, low and low, high and high, low and low. I'll hear a nice loud sound. Picture how dramatic this is. Two speakers blaring, there are spots where it is totally dead silent and other spots where it is loud. Then you keep moving. They'll go back out of synch again, and now at a different place in space it will be a steady, quiet tone. We call those nodes, and audiophiles are quite aware of this. They set up their speakers in their room, and they make sure there is some echoing going on, and they make sure that the speaker wires are not reversed on one of the speakers. If the two speakers are in phase, you can still find a node. You just have to shift away from the midpoint, and that's why a good audiophile will sit nice and equidistant between their well-designed speakers.

I can think of another experiment, which would also be pretty much of a clincher. Sound is supposed to be a pressure wave in a medium, so if you eliminate the medium, then there are no air molecules left to make a high-pressure region so the sound should go away. Whereas, if I believe in particles, if I believe that sound is really the flow of particles, then particles wouldn't care if they were going through a vacuum because there would be nothing to bump into. They would just cruise along.

This experiment was done back in the 1600s where somebody took a noisy object. Nowadays you can put a bell inside of a jar and you can see the clapper on the bell clapping and you can hear it. Then you pump the air out of the jar, and the less air there is the quieter it is. When you have vacuum, when there is no medium, then there is no sound. You can still see the little clapper clapping, but you can't hear a thing. Then you let the air back in again, and the sound comes back because you have some medium for the wave to propagate through.

I can never prove a scientific hypothesis. I can't prove to you that sound is a wave, but this kind of argumentation certainly goes a long way. I have proven from two very clear experiments that it's not particles. We have thrown away a whole class of models, and I've certainly shown that it's consistent with sound being a wave. I've made many predictions, and every time I make a prediction and I go out and test it, it comes true. That's the way nature is, and so this gives us great confidence in this model that sound is a wave even though we don't really physically see with a microscope anything going up and down. You don't see the air pressure. Air is invisible, and yet we know now that sound is a wave.

Let me think now about light. Light is an even more difficult story, and it's more difficult for a good reason. The good reason is the following. Waves have a wavelength. Sound waves—it depends on the pitch that you're listening to— but sound waves have kind of a typical wavelength of a human scale, a few centimeters, kind of the size of your ear, probably not by coincidence. Because of that, we can notice effects such as the one I described with the speakers. As you move from side to side, you can move your ear a distance that is comparable with one wavelength of sound. If you want to go from a place where the two waves are out of phase to a place where the two waves are in phase, you have to move basically one-half of a wavelength. If the wavelength is human scale, it's easy to do that, and you hear loud here, soft here, loud here—right? It's a pattern in the room.

With light, it turns out is that the wavelength is very, very small. It's less than a micrometer, and because of that, the slightest wiggle would move you from one peak to a trough and then to another peak and another trough. You would be washing out this effect. You wouldn't be able to stand at one place and obviously tell that. It wouldn't be silence now. It would be darkness, and then brightness, and then darkness, and then brightness. Think about how spectacular and weird that would be. If there were two light bulbs in front of you and you could move your eyes to some spot where all of the sudden they canceled out, that would be an amazing and direct proof that light is a wave and not light is a bunch of particles coming out of the light bulbs. You've never seen that with your ordinary eyes with ordinary light bulbs, and nobody had seen that back in history. That's why people were arguing so vehemently about whether light is particle or wave.

Isaac Newton believed that light is made of light particles, that a light bulb is emanating little light particles in all directions, and it's not a bad hypothesis. I can give you some arguments why you might want to believe that. Remember, I said that waves bend around corners. The evidence for that was that you can hear a sound behind a door even though you can't see the sound maker. However, you can't see a person behind the door unless they're straight through the door. The light is not bending around the corner. That seems to be evidence that light consists of a bunch of little particles. Now, it is also true about waves, that the smaller the wavelength, the less they tend to bend around corners. You can check this with water waves of different wavelengths. The bigger the wavelength is compared to the opening, the more bending you tend to have. It's not proof that light is particles, but there was no evidence back in Newtonian days that light is a wave.

Isaac Newton was thinking about other properties. He was thinking about color. He was thinking about the path of light rays through optical elements, and so this argument was kind of a philosophical argument for him. He didn't have the experimental equipment required to show convincingly, once and for all that light is a wave or particles. Although he had his beliefs and articulated them, they didn't really matter for the science that he was doing. It was in 1801 that Thomas Young did this dramatic experiment that I just described where you have two light bulbs and you can find a spot where they cancel one another. Young showed to the world that light really is a wave. It's very much like that speaker experiment, and what he needed was very high precision. It was a challenging experiment.

Here is the basic idea. Young took a very, very bright light source, then he ran it into a black wall in which he had cut a very, very narrow little slit—one slit to let the light through. When you let light through a very narrow slit, if it's narrow enough that it goes down to the distance scale of the wavelength of light, then the light will come through the slit just like water going through a little slit. It looks like you have a point source, and the light will come out in all directions. Now you have the equivalent of a point-like source of light, and if you were to look at that slit, you would see this little point-like source of light. The light is going out in all directions. You can tell because it's illuminating the wall on the far side of the room uniformly, and it does become a little bit dimmer and dimmer as you move farther away, but that makes total sense whether you believe that light is a bunch of particles or that light is a wave.

So far, no proof either way. Light could be little particles going through the slit bouncing every which way like particles bouncing off of edges, scattering and going in all sorts of directions. We haven't yet proven that light is a wave, but now on the back wall, that's uniformly illuminated, let's put two slits. It's called Young's two-slit experiment for this reason. Now think about the light that's striking those two slits. They are symmetrical with respect to the first slit. There are wave fronts heading outward and where nowadays I would think of the electric field, when the electric field is high, at one point if you go anywhere along the wave front, the electric field will be high, and then low, and then high, and then low. I'm visualizing this as a water wave where the high spot is expanding in a nice, beautiful, spherical pattern reaching the two slits symmetrically so these two slits are now in phase. When one is up, the other is up. When one is down, the other is down. Now, Young has no idea what's moving. He has this sort of vague idea that it might be a wave, and if it's a wave, he thinks something is moving. We'll come back to that point. Whatever it is, waves go positive and negative, positive and negative, and so now you have two light sources, two slits, which are coherent. They are in sync or in phase.

Now go on to the far side of the two slits. Turn around and look back at them. You see two little bright spots. Let's consider the two hypotheses. If light is particles, then you have little particles spewing out of one hole and little particles spewing out of the second hole. It's back to my argument that if you have particles coming at you from two places you're going to see them no matter where you are. As you move farther and farther away, it will become dimmer and dimmer, but basically you expect a pattern that's brightest right behind the slit and then fading off smoothly as you move off toward the edges. That's what I would expect if I believe that light is particles coming through the slits. If I believe that light is a wave and that waves are coming out of the slits and they're starting off in sync, then there should be places back on that back wall where, for instance, if it's equidistant to the two slits, then they both started in sync and they are waving toward me and still in sync. So I will see a bright spot, but if I move off to the side a little bit, if I move just far enough that one of the waves has gone half a wave farther than the other, half a wave means when one is up the other is down and vice versa—down, up, up, down, down, up, up down. They are canceling. They are destructively interfering with one another, and there's darkness. With my eye, I would be looking at these two bright slits. Then I move over a little bit, and now I can't see the slits. Then I move over a little bit further, and I see them again. It's remarkable. It's hard to imagine, but it's absolutely correct. That's what you see in the laboratory.

You could take a piece of film and develop it, and you'll see a bright band, then a pitch black band, and then a bright band. It's called an interference pattern, and it is as clear a signal that you have a wave as anything I can think of. It's very, very dramatic, and the pattern is completely predictable from the wave model. I can tell you how far apart bright and dark should be. It's geometry. When do the waves add up? When do the waves cancel? I can also predict what happens if I change the color of the light because that would change the frequency of the light and therefore the wavelength. I should be able to make a concrete mathematical prediction about how the pattern spreads out or narrows a little bit if I change the color, or if I change the distance between the slits, or if I change the thickness of the slits—many, many experiments all of which are completely compatible with a wave model for light.

As you think about all of these extensions and consequences, you realize that the wave model of light is convincing to the point that you accept it as the truth about nature. That's what we mean when we talk about scientific truth, is that in experiment after experiment, new situations and completely novel situations, they all match up with this one simple hypothesis.

By 1801, people know that light is a wave. Newton was wrong. That's a big deal and very exciting for Mr. Young to be able to show that Isaac Newton is wrong about something. Physicists love to be able to show that one of our great heroes has made mistakes, because this is the way science progresses, and as every revolution in physics, this one took a little bit of time. It was written up by Young in a way that some people found a little bit hard to understand, a little bit hard to believe. It is a pretty dramatic experiment, and so other people had to repeat the experiment. These verifications and extensions had to be checked, but it was a pretty quick revolution at which point there was this huge puzzle. It's clear to everybody that light is a wave, but what is waving? As we talked about over the last few lectures, it took another 60 years until Maxwell in the 1860's demonstrated mathematically that light is indeed a wave. It's an electromagnetic wave. It's not some material substance that is lifting up or moving back and forth. It's not a physical motion. It's electric and magnetic fields, which are waving in strength, stronger, weaker, stronger, weaker.

Once you have understood that light is a wave and sound is a wave, you can start to take advantage of it. I mentioned those sound-canceling headphones, a lovely little invention. Once you know that sound is a wave, here is this clever idea. Suppose you're in an airplane, and there is a drone from the engines. It's a very steady, low-frequency sound wave striking your ear. You build a little headphone that produces another sound. Those sound-canceling headphones are producing a loud sound, as loud as the jet airplane at the location of your ear. It's a very loud sound they make, but they're monitoring the jet sound, and when the jet pressure wave is high, they emanate a sound wave, which has a very pressure, and vice versa. They cancel the outside noise by adding their own interfering noise, and it's quite spectacular that you can do that. They work pretty well. They work best for steady, outside drones that aren't changing with time so that you can steadily create your own wave that cancels exactly with the outside wave. You can do this with light as well, anti-reflective coating is just a simple design that takes advantage of the fact that light comes and bounces off the front or the back. Both ways you're going to have some sort of interference of the two different paths, and you can make that interference cancel out light, which is in the middle of the visible range, and so you won't have a strong reflection.

The wave model allows us to not only understand the nature of these things, light and sound, but it also allows us to build devices and design new experiments. In the long run, it's very powerful to be able to separate a wave from a particle, and it helps us to understand how we think about different things in the world much, much more clearly.

Lecture Twenty-One
The Atomic Hypothesis

If, in some cataclysm, all scientific knowledge were to be destroyed, and only one sentence passed on to the next generation of creatures, what statement would contain the most information in the fewest words? I believe it is the atomic hypothesis (or atomic fact, or whatever you wish to call it) that all things are made of atoms—little particles that move around in perpetual motion, attracting each other when they are a little distance apart, but repelling upon being squeezed into one another. In that one sentence you will see an enormous amount of information about the world, if just a little imagination and thinking are applied.

—Richard Feynman, from the *Feynman Lectures in Physics*, vol. I

Scope:

In the first part of this course, we looked at fundamental, underlying principles of physics, explored particular forces, and saw that electricity and magnetism were unified and that the resulting force—electromagnetism—helps us to understand light and light waves. There's one more important part of the story—the idea that the world is made of atoms, independent, fundamental building blocks of matter. We'll trace this idea's long and complex history, which offers a unifying principle greater than Maxwell's equations. Energy, the structure of materials, chemistry, heat, optics, and more become tied together and often relatively simple to describe and explain at a fundamental level, once we have a basic understanding of atoms.

Outline

I. Greek natural philosophers debated what the world is made of.

A. In about 400 B.C., Democritus (c. 470–380 B.C.E.) promoted a fairly modern idea of atoms: that at a fundamental level, the world is made of "uncuttable" (*atomos*) objects.

B. Aristotle (384–322 B.C.E.) disagreed; his worldview encompassed qualities, which would be infinitely divisible.

C. At the time, this was a philosophical, not a scientific, debate;

people didn't consider measurable consequences of one idea versus the other.

1. A stick of butter is a material that has certain defining characteristics: it's yellow, it melts at room temperature, and so on. If I cut the stick of butter in half, it's still butter; I haven't changed the essential character of it. I could continue this process of cutting the butter in half, and all along, I would still have butter.
2. Democritus argued that at some point, I would reach an extremely tiny piece of butter that I could no longer cut. As we said, Aristotle disagreed with this idea. (For him, it's "butter all the way")

D. To decide this question, we must consider the consequences of both theories.

II. Atoms are observable with modern equipment, such as a scanning electron microscope. Even in the 1700s, however, people were becoming convinced that the idea of atoms was a useful and correct description of nature.

A. An atomic worldview is quite consistent with the classical physics ideas of reductionism and determinism, yet it also forms a bridge to modern ideas. In fact, I would argue that the classical physics story ends with the atomic theory; modern physics then takes over, looking at what the atom itself is made of. Even without the atomic worldview, much of classical physics still "works."

B. According to the atomic hypothesis, there are about 100 different kinds of atoms, such as carbon, nitrogen, oxygen, hydrogen, and so on, that combine to form all material objects—solids, liquids, and gases.

C. If we know the mass of the atoms and their interactions with other atoms, we can build (and understand) any material substance.

III. The idea of atoms helps us make sense of both physics and chemistry.

A. In Newton's era, alchemy served as a sort of proto-chemistry. Alchemy involved experimentation with materials, but it was practiced in secret and had a mystical aura.

B. Antoine Lavoisier (1743–1794), a French chemist working at the end of the 18th century, took the first classically scientific steps in chemistry. He showed, for example, that mass is conserved in chemical reactions, which tended to confirm the atomic hypothesis.

C. John Dalton (1766–1844), a British chemist working soon after Lavoisier, is called the father of the atomic model. He carefully studied many aspects of chemistry. Let's look at one example: forming water.

1. Mixing 2 parts (by mass) of hydrogen with 16 parts (by mass) of oxygen yields 18 parts of water vapor.

2. Looking instead at volumes, the combining ratios are 2 volume elements of hydrogen and 1 volume element of oxygen to yield 1 volume of water vapor. This makes sense if water is H_2O, that is, if a water molecule has 2 parts hydrogen for every 1 part oxygen.

D. In the end, the atomic model is a simple and elegant system that allows us to organize the elements and make sense of their properties.

IV. The experimental and theoretical work of Robert Boyle (1627–1691) in the late 1600s, Jacques Charles (1746–1823) in the late 1700s, and Amedeo Avogadro (1776–1856) in the early 1800s followed a parallel path toward the atomic hypothesis, this time, from the perspective of physics.

A. Instead of combining elements and looking at ratios and masses, these investigators were doing physics, that is, measuring forces, pressure, temperature, and so on. They found that the atomic hypothesis of chemistry helped them make sense of the physical properties of gases.

B. All gases have some universal characteristics. For example, they obey the *universal gas law*, or *ideal gas law*, which involves a relationship of pressure, volume, and temperature.

C. The atomic hypothesis characterizes a gas as a collection of atoms, independent "superballs" flying around in space. Pressure, then, results from these objects bouncing off the walls. This is an extension of the atomic hypothesis, a branch of physics called *statistical mechanics* (used in the explanation of atomic systems).

V. The atomic model leads us to much deeper questions about nature.

A. What makes a solid substance melt? The atoms in a solid state are bound together by chemical forces. As the substance is heated, the atoms gain kinetic energy; when the energy reaches a certain point, the chemical bond is broken and the atoms are free to move around. The result is a liquid, but note that in a liquid, the atoms are still in contact with one another. If the substance is heated still more, the atoms reach an energy level at which they lose contact, becoming a gas.

B. What makes a substance dissolve? Sugar stirred into water seems to disappear—where did it go? The atoms making up the sugar migrate into the spaces between the water molecules.

C. These are qualitative questions about atoms, but we might also begin to ask quantitative questions. For example, how big are atoms?

1. To answer this question, Benjamin Franklin conducted an experiment in which he placed a drop of oil onto a pond. By equating the volume of the drop to the volume of the resulting oil slick, Franklin came up with an estimate of the height of the slick, very roughly 1 atomic diameter, about a billionth of a meter.

2. The size of an atom can also be measured by a technique called *X-ray interference* and is comparable to the wavelength of X-rays.

D. The atomic hypothesis also helps us understand temperature (a measure of the energy of atoms). We'll explore this idea in the next lecture.

VI. Atoms are somewhat abstract to us because no one has ever seen one and no one ever will (they are much, much smaller than the visible wavelength of light). By now, however, atoms are a well-established idea in physics. We can see the consequences of the atomic worldview all around us, enabling us to explain and calculate the properties of ordinary objects.

Essential Computer Sim:

Go to http://phet.colorado.edu and play with Gas Properties to study the ideal gas law and get a visual sense for how the atomic model is directly responsible for the observables, such as temperature and pressure. Balloons and Buoyancy may help you understand *why* a hot air balloon rises, based on the atomic model. For further investigation, check out any of the sims in the Chemistry category.

Essential Reading:

Hewitt, chapter 10.

Hobson, chapter 2.

March, chapter 13.

Recommended Reading:

Cropper, chapter 13.

Feynman Lectures, introductory chapter on atoms.

Questions to Consider:

1. What does the atomic hypothesis predict will happen to gas pressure as you increase its temperature? (Temperature measures the average kinetic energy of atoms, and pressure is related to the force the atoms apply to the walls. If you increase energy, what happens to the force atoms apply to the walls? What happens to the frequency at which atoms bounce off the walls?)
2. Which has more atoms, 1 kg of nitrogen gas or 1 kg of hydrogen gas? (Note: Hydrogen atoms are the lightest possible atoms. Nitrogen atoms are 14 times heavier than hydrogen.) Which has more atoms, 1 liter of nitrogen gas or 1 liter of hydrogen gas? (A liter is a measure of volume, not mass.)
3. Copper atoms have a mass of 63 *atomic mass units* (each *amu* is 1.66×10^{-27} grams). Estimate the mass of a penny. (A stamp scale can help. Otherwise, can you think of a way of comparing an unknown mass to known masses, for example, with a pile of pennies?) From that, estimate the number of atoms in a penny. Now estimate the volume of a penny (if you don't have a ruler to measure something as small in thickness as a penny, could you come up with a trick by stacking pennies to make an estimate?)

Given the total volume and your estimate of the number of atoms, what is the volume of one atom? Assuming that the atom is a little "cube," what is the size of the atom? (Your answer should come out to be around 10^{-9} meters on a side.) If you look at a periodic table (go to the Web!), you can find atomic masses of all elements. Make a similar estimate for other materials you find in your house. Is the size of all atoms about the same?

4. Thinking more about the previous question. How could you figure out the mass of a single copper atom (without looking it up in the periodic table)? In other words, how did people figure out the mass numbers in that table?

Lecture Twenty-One—Transcript
The Atomic Hypothesis

We spent the first part of this course thinking about the fundamental underlying principles of physics. We looked at kinematics, Newton's laws, in particular F=ma, the principle of conservation of momentum, and conservation of energy. Then, armed with that groundwork, we started looking at particular forces. We've looked at gravity. We've looked at electricity and then magnetism, and then we recognized that electricity and magnetism were unified. Maxwell helped us to visualize this unification and think about electricity and magnetism as flip sides of one fundamental true force of nature. We discovered that force of nature tells us about light and light waves, which led us to think about waves as a general feature of the world. That spread of ideas covers a good chunk of classical physics. There is one key important final story, which will lead off to many different consequences. It is perhaps one of the greatest ideas of all of physics, and I've been alluding to it all along. It's part of our culture, and everybody knows that the world is made of atoms. Nobody knew that. Even 100 years ago, people were arguing about it, and 300 years ago people had no idea about how you would demonstrate whether the world was made of atoms or not.

Today I want to talk about the idea of atoms, how we could convince ourselves that the world is made of atoms during the period of classical physics, from Newton all the way until contemporary times. Let's go back a little bit first and think about the Greek philosophers. People have always been wondering about what the world is made of. What are the building blocks? It's a very natural question. Four hundred years B.C.E., Democritus had some writings in which he postulated that the world is made of fundamental little "uncuttable" objects. It's the bottom of the line, and "uncuttable" in Greek is *atomos*. That's the origin of the idea of atoms. Aristotle looked at these ideas and said, no, I don't think so. Remember, Aristotle believes in qualities, and he believed that qualities are not divisible into chunks, and so he had a worldview that the world can be divided and subdivided indefinitely. In that era, it was a philosophical debate. People couldn't really think about measurable consequences of one idea versus the other to test so that you could disprove one and begin to convince yourself of the other.

Let's think about a picture of this. If I give you a stick of butter, butter is a material and has certain defining characteristics. Aristotle would be perfectly happy to start listing them. It's yellowish, and it melts at room temperature. It has a certain density. Density is mass per unit volume. That's not quite how heavy it is. It's how heavy it is for a given amount, and if I chop the butter in half and now I just investigate half a stick of butter, it's still clearly butter. I haven't changed the character of the material. I have less of it, but it's still yellow and it still melts at the same temperature. It still has the same mass per volume. I have half as much mass, but I also have half the volume. You can now start chopping it in half and in half. Now you have a patient. Now you have a fraction of a pat, but it's still butter. What Democritus argued was, it's not butter all the way. There's one final cut, and he didn't know where it was, at what point or how big that last piece was. He assumed that it was extremely tiny but not infinitely tiny, and that's the last piece of butter. If you try cutting that, you can't. If you did, you wouldn't have butter anymore. Whereas, Aristotle argued, it's butter all the way. You could just keep dividing it in half and in half as many times as you like. It depends on the sharpness of your knife.

How are we going to decide this question? Well, the nature of science, classical physics, is to think about consequences. If you believe that the world is made of atoms, what measurable consequences can you think up and we'll go test those.

We're trying to create a worldview that's as simple as possible. That's one compelling argument for why you might believe in atoms, that is because it's a nice simple point of view, but it has to go beyond that and we have to be able to test this. People spend entire lifetimes trying to think about the consequence of the atomic hypothesis, and I would like to sketch out some of the big ideas.

Nowadays we can image atoms not with our eyeballs, but with devices like scanning electron microscopes. We can see individual atoms by the evidence left behind on a computer screen. So we certainly have very direct evidence today that atoms exist. They're real, but even back in the 1700's there were already some pretty good clues. People were beginning to develop the idea and becoming more and more convinced that it seemed like a useful and ultimately correct description of nature.

It turns out that the atom is a lovely topic in classical physics. I think atoms belong squarely in the field of classical physics, and yet they also form a bridge to modern ideas because classical physicists want to know what the world is made of. We've been pretending that the world is made of point-like objects, and now we're talking about what those point-like objects are. Today we ask what the atoms themselves are made of. You can keep on digging deeper and deeper, and I would argue that the classical story ends with the atom. The atom is the bottom line. It's the simple, "uncuttable", fundamental building block, and so that's where we're going to zoom in for this lecture.

If you don't believe in atoms, or you don't know about atoms, much of classical physics still continues to work. You can talk about Newton's law for the moon or for a baseball, and we can pretend that the baseball is a point-like object or that the moon, for that matter, is a point-like object. That works just fine. We've talked about this throughout the course—that we can simplify our thinking—but now we're asking, well, are there true point-like objects? Is this a physical reality? If there are, then our philosophical vantage point that the world is ultimately simple, our idea of reductionism, that complex things can be broken down ultimately into their building blocks and determinism, that if you know the building blocks and you know Newton's laws and conservation laws you can build up to arbitrarily complex systems, All of this leads us to believe that the idea of atoms could be very fruitful.

Let me be a little bit more articulate about what the atomic hypothesis says. The world is made of atoms. Every material object, solids, liquids, gases—all objects are made of fundamental atoms. There are different kinds of atoms. For instance, carbon is one kind of atom. Nitrogen is another. Oxygen is another and hydrogen yet another. These are all atoms that are present in our bodies. Those four kinds of atoms describe a good chunk of your body, and if you add a few more, they describe a good chunk of the physical world. All you're doing is combining them in different ways.

In the end, it turns out that we need about a hundred of these things to describe everything we've ever observed, just a few more than a hundred fundamental atoms. And every atom, of course, can be repeated so you can have many different carbon atoms, but they're all identical, absolutely indistinguishable from one another and that's what the world is built of.

The periodic table, the graph of the elements that is always up on the blackboard or above the blackboard in chemistry and physics classes, is a list of all of the known atoms. Hydrogen is in the upper left-hand corner and then helium, working its way down through the periodic table listing all of these hundred.

If you know them and you understand their properties, you understand what they're made of. All you really need to understand is that they have a mass, and they have certain interactions with one another. Then you can build anything else. What is butter? Butter is made of molecules, which are built out of carbon, nitrogen, oxygen and hydrogen. The idea of atoms is going to help us. It's going to help us to make sense of physics and chemistry and many other branches of physics and chemistry in particular.

At the beginning of this course, I was a little bit dismissive of chemistry. I argued that it is derived from physics, and so physics is the fundamental. Chemists, of course, would bristle at that thought, and I have to give credit where credit is due. The idea of atoms really arose from chemistry. Chemistry and physics at one point, back in the 1700's, were intimately connected. They were investigations into the world, the natural world we live in, and the way they were investigating it evolved until nowadays, people who mix chemicals and look at the reactions are doing chemistry. We don't call that physics anymore, but it really still is trying to understand the building blocks of the world in many respects.

Now, back in Newton's era there was no chemistry. Of course, there was no physics either. He invented it, but Newton spent time doing what we now would call the beginning of physics. He also spent at least as much time, maybe more, doing what we call alchemy. Now, it has "chemy" in it; it sounds as if it is related to chemistry. It's kind of a proto-chemistry. It's a mix of mysticism, superstition and a quasi-religious cult approach to understanding the world. Nowadays, at least as far as I know, there is nobody seriously pursuing alchemy, but in Newton's era doing alchemy meant what you think of as classic chemistry experiments—mixing things, heating them up, grinding different materials and trying to figure out what's going on, except that alchemists were not behaving like scientists in the contemporary sense. So in particular, the culture Newton was helping to create in physics that was so enormously valuable, is what it means to be a physicist—to do experiments, to try to hypothesize about an underlying physical explanation, to then publish this work, to go to meetings to talk to other knowledgeable but skeptical scientists who are interested, who can themselves repeat your experiments and figure out where you're right and where you're wrong, and come up with alternative or improved hypotheses—that whole culture of science didn't surround alchemy. Alchemy was done by yourself, and you would write up your results in little cryptic codes so that other people couldn't steal what you did. It was by and large recipes. It wasn't a fundamental, principle-based scientific study. Because of that, Isaac Newton—although he spent an enormous amount of time doing alchemy and he wrote quite a bit about his alchemy—very little of his work followed through and contributed to our contemporary scientific development. It's too bad, but if he had treated alchemy in the same way that he had treated optics, mechanics, or astronomy, it's quite possible we would have progressed more rapidly in the chemistry field.

I want to talk about how we started developing the atomic model by following two paths, and one of the paths was with the chemist. The other path was with the physicist. I would like to start with some chemists and think about the early progression so that we can understand why you might have believed in this crazy idea back in an era, when there was no hope at that time of being able to see individual atoms.

Antoine Lavoisier, a French chemist, working roughly 100 years after Isaac Newton, at which point chemistry had become something of a science, really helped to firm it up. He makes very, very careful measurements, and he looks at, what if you mix this material in some proportion with that other material in some proportion. He noticed, for example, that if you mix material number one and you have a certain mass, and you mix it with material two and you have a certain mass that the total mass of the product you have is conserved, though it might be a new material and have new properties—because when you do chemistry, you can change the character of what you're working with, because a chemical reaction has occurred. This is a big idea because it makes you think, why should that be? How can we make sense of that experimental fact at the macro level? A micro explanation could be that we are just working with little tinker toys.

You take one piece of mass from this one, one piece of mass from that one, you add them, they hook together, and you still have the same amount of mass. They're just hooked together now. That tends to make you think that the materials you're working with are built of little building blocks. If it had continuous properties, then it would be very difficult to make sense of the observation that mass is conserved.

Soon after Lavoisier comes John Dalton. He is a British chemist, and he's called the Father of the Atomic Model. He really wrote it down, talked about it, and thought about why you might believe that the world is made of these building blocks. Not only does he do these mass measurements, but he does very careful measurements of many aspects of chemistry. Let's think about one in particular. Think about forming water. You start with hydrogen gas, and you have some oxygen gas. These come in tanks with a measured mass and a measured volume, and you combine them together. You might need a little spark to make it go, and you have water. If everything is hot, you'll have water vapor. You have a different kind of gas coming out. If you look at the masses involved, you will discover that you have sixteen parts by mass of oxygen combining with two parts by mass of hydrogen yielding sixteen plus two, conservation of mass is eighteen parts by mass of water vapor.

How could we understand that? Well, Dalton is saying that maybe what is going on is that we are combining a certain number of hydrogen atoms and a certain number of oxygen atoms to form water. Well, you could call them atoms, but people call them molecules because we've combined atoms. Now, if you look at the volumes, you will have two volume elements of hydrogen combining with one volume element of oxygen producing one volume element of water vapor. That's not like conservation of mass. Two plus one equaled one. In the alchemy days, that would just be some arcane rule that people may or may not have noticed or cared about, but it didn't make any sense. Dalton says, this makes total sense if water is H2O—if a water molecule has two parts of hydrogen for every part of oxygen, then two parts by volume of hydrogen plus one part by volume of oxygen will form one water molecule, so one volume element of H2O. It's very, very simple, and all you need is to know some basic properties. For instance, now we conclude, since water is H2O, that a single oxygen atom must weigh sixteen times as much as a single hydrogen atom. That's how you the masses add up.

The hypothesis that the world is made of building blocks and now we know the mass of hydrogen and oxygen relative to one another and can do some more chemistry and figure out the relative masses of all of the different elements, we can start to learn their properties. Some of them are reactive in some way. We can begin to organize this periodic table of the elements, and in the end it's a very simple and elegant system that we built up that doesn't have very many fundamental constituents, and yet all of chemistry, and for that matter, presumably all of alchemy if we cared about that any more, could be understood just by this basic combination and recombination of atoms.

By this period in history, by the early 1800's, people are beginning to agree that atoms are a good idea. They're useful bookkeeping tools for understanding chemistry. Chemistry is no longer just rules. It's now principles. There's an underlying story that we can make sense of. You can still argue whether these little building blocks are real. Are they physical, or is this just some kind of mathematical bookkeeping game? People always worry about that when you invent a new idea of something that's invisible.

Let's shift from the chemists to the physicists. They are working in parallel. Back in the late 1600's, Robert Boyle was doing both experimental work with gases and some theoretical work, followed up by Jacques Charles in the late 1700's, and Amedeo Avogadro who was pulling the big story together in the early 1800's. These are following a different path. Instead of combining things and looking at the ratios and the masses, they are doing physics. Physics means you measure masses, you measure forces, you squeeze on the box of gas, you have some piston, and you make physical measurements. What is the temperature? What is the volume? What is the force per unit area? We call that the pressure. Pressure is just good old Newtonian force divided by the area that you've spread that force over. These folks were beginning to agree with the chemists. They said, yes, that hypothesis helps us to make sense of the properties, the physical properties of gases. What they were observing was that all gases have certain common behaviors, the universal gas law. The gas law tells you what happens when you squeeze a piston. The pressure and the temperature will change depending on the change in volume, and these relationships between these physical, measurable quantities were very, very regular. It worked for carbon dioxide, oxygen or hydrogen gas. It didn't really matter. The ideal gas law is how it is referred to today, and it refers to an ideal situation. Most gases that they were working with were very, very close to doing the same ideal behaviors and have the same relationships one with the other.

They didn't develop the model that I'm about to describe, the model that I described last lecture as well, but it did arise form this atomic hypothesis. If the world is made of atoms, what is a gas? Well, the gas is made of atoms. It's that little super ball story that the gas is really just a bunch of little super balls flying around. Oxygen gas is just a bunch of little oxygen molecules, little tiny compact Newtonian fundamental particles flying through the air, bumping into one another and bumping into the walls. What would pressure be in this model? Well, if you have little super balls bouncing against the wall at a given rate and they have a known speed distribution, you can conclude what the force would be. It's just Newton's law, F=ma. If you have little objects whacking into the wall, then you can calculate what the pressure would be as a function of the energy of those little objects, which is what temperature is measuring.

This is an extension of the atomic hypothesis. It's a branch of physics that really takes the atomic hypothesis seriously, and we now call it statistical mechanics—mechanics, because like classical mechanics that we've been studying all along, it's the explanation of atomic systems. Any kind of a gas is really built up out of atoms, and statistical means that we have an awful lot of them. These little super balls are extraordinarily tiny so we're not going to worry about the motion of one or the motion of the other. We're not going to try to track them. We're going to look at averages, and we're going to understand pressure and volume as just having bulk consequences arising from this microscopic model.

The atomic hypothesis goes much further than just the ideal gas law. This is really the starting point, and once you have the building blocks and it is consistent between the chemistry story and the ideal gas law story, you're beginning to ask yourself deep questions, such as, if this is the way the world is, we should be able to understand everything, every object, not just these ideal gases, but liquids and solids. You could ask questions such as, what makes an object melt. A solid object becomes hotter. How do we make sense of that? Well, the atoms, because we believe in atoms now, must have been in lock step. They were in a solid so they were bound together by chemical forces, and as you heat it up, what are you doing? Well, you're making the little atoms jitter more and more rapidly. It's simple Newtonian physics. You're giving energy to the little atoms, kinetic energy, energy of motion, and as they jitter more and more rapidly, there will come a point when they break the chemical bond. There's a certain potential energy associated with the little chemical bonds. It's all straight classical physics, and at that point, the molecules would be free to roam around. That's what a liquid would look like. Instead of all being locked together and having a solid, they are now free to roam around. They have now separated one from the other, and that would be a liquid where they're still in contact with one another but no longer rigidly locked.

If you keep heating it, then ultimately they will fly up into the air because they have so much energy. They're now high-energy super balls bouncing around the room, and that would be evaporation. You can understand all sorts of physical events—evaporation, sublimation, and melting—all in terms of the atomic hypothesis. You could understand dissolving. Take some sugar and put it in water. Why does the sugar disappear? Why does the volume of the water not change in a very dramatic way even though you put a big volume of sugar in there? Where did it go? The mass is still there, but the volume didn't change. Well, the atomic hypothesis helps us to make sense of this because the sugar is built up of little atoms. Those atoms can react with the water and begin to migrate in between the water molecules. There is space in a liquid between the water molecules, and so the pieces of sugar, the atoms that are making up the sugar or at least combinations of those atoms, can fit into the interstitial spots. You can speak qualitatively about all of these things, but you can also start to talk quantitatively about all of these things.

For instance, you could ask, how big are these atoms? That was a question that Democritus had no way of tackling, but by the 1700's and 1800's people were starting to come up with ways of determining the size of atoms. Ben Franklin came up with a very clever little experiment. He took a drop of oil. He knew the volume of that drop. It's just a little sphere, and he dropped it on top of a pond. It spread out and made an oil slick. It spread out, spread out, and then it stopped spreading. It had a certain size to it, and he could easily measure the area. What he couldn't do was measure the thickness. It was so thin that no possible ruler in his era could measure the thickness, but by believing that the oil is still oil, that the oil is made of atoms that have just spread out until you only have perhaps one or two atoms thick because they can't spread any thinner than that, then you can deduce that the volume that you started with should be the volume that you ended with because it's the same material. It's just a different physical configuration. The volume of a liquid doesn't change as you change the shape of the container.

Ben Franklin measured the area and deduced the height, and he came up with one of the first estimates that turns out to be very good, quantitatively, of the typical size of an atom, about a billionth of a meter, one/one-billionth of a meter. That's how thick that oil slick is—well, it's a few billionths of a meter. Then there were many, many, many experiments to come, and in each of these experiments, very, very different kinds of experiments, the size of the atoms always seemed to come out to be about the same size. Atoms seemed to have a characteristic size, which again leads you to believe that they are really fundamental objects in the world.

One of the ways that you can determine the size of an atom would be to shine X-rays at it. We talked about how X-rays like any form of electromagnetic radiation can interfere if they're running through slits and if the slit size is comparable to the wavelength of the light. Then you'll have a beautiful interference pattern, and so with the X-rays people recognized that the wavelength of the X-rays was comparable to the size of the atoms themselves. The atoms were the slits in a crystal. Many, many different kinds of experiments all came at the same fundamental physical picture. If you think about heat and temperature and now the statistical mechanics, this thinking about the world as being built up out of little atoms is beginning to make more and more sense. As you heat something up, the atoms jiggle more rapidly, and now we realize what temperature is. What is it measuring? It's measuring the jiggle rate. It's measuring the energy of those little molecules. When you ask questions about what happens when you heat something, this atomic worldview will help us to make sense of what's going on.

That's what we're going to cover in our next lecture, as we begin to think about further consequences of the atomic hypothesis. I recognize that atoms can be an abstract idea because they're so small that nobody has ever seen one, and nobody ever will. They are much, much smaller than the wavelength of visible light, and if an object is much smaller than the wavelength of light, the light won't interact in a useful way with it such that you can directly image it. You cannot and will not see an atom. You can only see it indirectly by using something other than visible light. You'll need some kind of detector such as an electron microscope detector.

By now atoms are as well established an idea in physics as I can think of, and there are many, many consequences. When you look at the world around you and you think about the properties of ordinary objects, properties such as thermal properties, optical properties, mechanical properties, the color of the object—which is really just the interaction of the light with the electromagnetic radiation—anything you can think of, really, can be made sense of, and maybe even calculated on the basis of this underlying atomic hypothesis.

Lecture Twenty-Two

Energy in Systems—Heat and Thermodynamics

Thermodynamics is the only physical theory of universal content which, within the framework of the applicability of its basic concepts, I am convinced will never be overthrown.

—Albert Einstein.

Scope:

So far in these lectures, we've tried to simplify as much and as often as possible, before adding complexity back in by degree. Historically, this strategy has been extremely productive in physics and continues to be used to this day. In this lecture and the remaining ones in the course, we'll make the transition from simplicity to the recognition that the world is constructed of atoms, and thus, even simple things may be enormously complicated. We'll look at the field of thermodynamics, the study of heat and temperature, which requires an understanding of microscopic internal degrees of freedom exhibited by atoms. The physics of thermodynamics is everywhere– we use it in heating and cooling our homes, cooking food, taking the temperature of a child with a fever, and predicting the melting of glaciers. Thermodynamics is an appropriate ending topic for this course because it pulls together all the "big ideas" of classical physics, including Newton's force laws, energy principles, the atomic hypothesis, and statistical mechanics.

Outline

I. Thermodynamics begins with a focus on energy flow using the principles of statistical mechanics (which amounts to "averaging over atoms").

- **A.** Complex, real-life systems involve astronomical numbers of particles. A pot of boiling water might contain a million billion billion molecules.
- **B.** The reductionist viewpoint tells us to focus in on individual particles and track their reactions with regard to Newton's laws. This would be a hopeless task (given such a large number of particles to track).
- **C.** Instead, statistical mechanics tells us to think about

averages—the behavior of *typical* atoms—without worrying about the details. This simplifies the story enormously.

D. Insurance companies use this principle when they make predictions about when people will marry, have children, or die. It's not possible to make such predictions about individuals, but it is possible to do so for a large pool of people.

E. Because atoms are simpler than people, predictions about average quantities are all the more reliable.

II. Thermodynamics is characterized by three laws, plus a starting (*zeroth*) law. The zeroth and first laws, which we'll look at in this lecture, are about work and energy, along with energy flow. The second and third laws add a new concept that we'll talk about in the next lecture—*entropy*.

III. The history of thermodynamics dates back to antiquity.

A. Early ideas about heat included the *caloric theory*, in which heat was a material substance (a physical fluid) associated with high temperatures.

B. The basis for the modern model of thermodynamics came from James Joule, whom we discussed in an earlier lecture. He articulated the idea that heat is the flow of thermal energy.

1. Recall that energy comes in many forms—kinetic energy, gravitational potential energy, chemical potential energy, and so on.
2. Think about a book sliding across a table. It starts with pure kinetic energy—energy of motion.
3. When friction grinds the book to a halt, where did the energy go? It is not stored in an obvious way, as you might see in a compressed spring.
4. With the atomic hypothesis, we know exactly what happened to the energy. The friction caused an increase in the motion of atoms in the book and table; thus, they have more kinetic energy. The thermal energy is stored in random kinetic energy of these atoms.
5. Thermal energy is measured in joules, just like any other kind of energy; we could measure quantitatively the

thermal energy of the book and the table.

C. Experiments in thermodynamics were difficult to conduct, and progress in this field of physics was slow.

1. These ideas might have been accessible to Newton, but it took almost 200 years to put the story together carefully.
2. Part of the problem was that it's difficult to isolate a system thermally. It's also difficult to measure small temperature changes accurately, and older thermometers tended to interfere with scientific analysis.

D. Many physicists worked for years to make a convincing case that thermal energy is, indeed, just another form of energy, not some mysterious caloric fluid.

1. We can think of some simplistic arguments against the caloric theory. For example, if you have a cup of coffee that is heated by the "caloric," it should weigh less as it cools off and the caloric leaves it.
2. Suppose you have a block of dense material, and you want to drill through it. You know what will happen: As the drill bit grinds away, it will get hot. But where is the caloric coming from in this system? Is it created out of nothing by the interaction of two cool objects?
3. Ultimately, many experiments verified Joule's idea that heat was not a material substance but related directly to energy.

IV. The zeroth law of thermodynamics defines *thermal equilibrium*.

A. Two objects in contact with each other may or may not be in thermal equilibrium. If they're not, they will change; one of them will cool off and one of them will warm up until nothing more happens. The two objects are then in thermal equilibrium.

B. The zeroth law of thermodynamics says that if object A is in equilibrium with object B, and object B is in equilibrium with object C, then A is in equilibrium with C. This is a practical statement, allowing us to use thermometers reliably.

C. Temperature becomes meaningful with the zeroth law of thermodynamics, and the law tells us that we measure temperature by comparison with a standard.

D. Microscopically, the zeroth law also makes sense. Temperature measures the average kinetic energy of the atoms in a system. The atoms in my body have a certain average kinetic energy; when I take my temperature, the thermometer reaches thermal equilibrium with my body.

1. If the thermometer starts off cooler than my body temperature, its atoms are moving more slowly.
2. What happens if we bring two solid objects into contact, one of them with atoms moving slowly and one of them with atoms moving rapidly? The atoms that are moving rapidly will bump into the slow ones more frequently and speed them up. Of course, the atoms that were moving rapidly will also slow down in the process.
3. In the end, in equilibrium, the *average* energy of all the atoms will be the same.
4. Temperature has nothing to do with the material object involved; it is nothing more or less than the average kinetic energy of atoms.

V. The first law of thermodynamics is a statement of energy conservation for complex objects.

A. To understand this idea, we need to define our vocabulary: *temperature*, *thermal energy*, and *heat*.

1. *Temperature* is a measure of the average kinetic energy of particles.
2. *Thermal energy* refers to the sum, not the average, of internal kinetic energies.
3. In physics, *heat* is used as a verb, not a noun. Heat is defined as the transfer of thermal energy from one object to another.

B. The first law of thermodynamics says that energy is conserved.

1. Total change in thermal energy arises from work plus heat.
2. If you put a pot of water on a hot stove the water will get hotter. That means that the average energy of atoms is increasing; where is this energy coming from?
3. A classic Newtonian concept is that energy is transferred

through work. Thus, *stirring* the water would be one way of increasing its temperature

4. In our scenario, though, we are transferring random motion of the atoms on the stovetop and converting that to random motion of atoms in the water. (So here, we increase the temperature of the water by heating, *rather* than by doing mechanical work.)
5. Joule argued that heat and mechanical work are equivalent; they can both be measured in the same way. In fact, he measured the *mechanical equivalent of heat.*

C. The bottom line so far is that objects can hold thermal energy (hidden, *internal energy*), but this energy is fundamentally no different than any other kind. Up to this point, thermodynamics is a bookkeeping tool that allows us to keep track of energy. In the next lecture, we'll talk about the entropy concept, which will take us beyond bookkeeping.

Essential Computer Sim:

Go to http://phet.colorado.edu and play with Gas Properties again. Click on the Energy Histograms box and see if you can make sense of the resulting graphs!

Essential Reading:

Hewitt, first half of chapter 17.

Hobson, start of chapter 7.

Recommended Reading:

Hewitt, chapters 13–16.

Cropper, chapters 6–8.

Questions to Consider:

1. A metal and a wooden object sit in the same room for a long time. Which one has the higher temperature, or are they the same? Why? In which one do the atoms have a higher average kinetic energy, or are they the same? Why? Now touch them; the metal one will *feel* cooler. Can you make sense of this, given your (probably correct) answer to the previous question?

2. What happens to the work done when you vigorously shake the orange juice you're mixing up?
3. When you put an ice cube in hot water, does temperature "flow" between the ice and the water? (If not, what does flow between them?)
4. Does a ceiling fan cool the air in a room? If not, why do you use one? What is it doing?
5. Use the principle of conservation of energy to explain why the temperature of the air in a bike pump increases when you compress it, but the temperature of compressed gas in a can decreases when you let the gas suddenly expand.
6. Can you convert internal (thermal) energy into useful (mechanical, kinetic) energy? If so, give some examples.
7. Use the first law of thermodynamics to explain why the total energy of an *isolated* system never changes. Does that mean that "nothing interesting" can ever happen to an isolated system?

Lecture Twenty-Two—Transcript
Energy in Systems—Heat and Thermodynamics

So far we've tried to simplify as much as possible, whenever possible. We treated objects like points. We considered a spherical cow. Even an athlete or an automobile we've thought of as a point. You focus your attention on the center of mass. You watch it move. It obeys Newton's laws, and then, of course, we recognize that the world is more complicated than that, and so we started adding in the complexities. You don't have to always neglect friction. Once you understand the underpinning laws of nature, friction is just another force. You can calculate it. You can observe it, and you can add it back in to the F side of F=ma. If you watch the athlete diving of the high dive, you can watch their center of mass, but you can also recognize that they're doing spins and turns, and you can calculate and understand the rotations around the center of mass or even the change in the shape of their body, which makes the story even more complicated. Fundamentally, we've been trying to treat objects as though they were as simple as we possibly could. That's why we were able to function historically for so long without worrying about whether atoms are real or not because we weren't really looking down all the way, very often, to this microscopic building-block level.

This is a very, very successful strategy. It has been extraordinarily productive throughout history and continues today. An astronomer even today can think about the motion of the Earth and understand the seasons and the phases of the moon, and you don't have to worry about such things as, well, there's Mount Everest and so it's not really a perfect sphere. Or we have an atmosphere so it's not really a solid body. It depends on the questions that you're asking. If you're worried about global warming, then you do have to worry about the atmosphere, and that little thin layer can have many important consequences.

In this lecture, and basically throughout the rest of the course, I want to make the transition from thinking about the world as simple as possible to recognizing that the world is built out of many little tiny atoms. Things, even simple things, are enormously complicated. Even granting this, it turns out that we can make sense of what's going on. It's the last piece of the classical physics story where we accept this reality and look for the consequences and again try to think about the consequences in as simple a way as we possibly can.

This is the field of thermodynamics. That's the name for the branch of physics we're talking about. It belongs to classical physics. Thermodynamics—thermo makes you think of temperature thermometers, and dynamics makes you think about why questions. Why does one object cool off and another object warm up? Why does heat flow the way it does? These are the kinds of questions that we want to investigate. When you study thermodynamics, you begin to recognize that the microscopic, internal degrees of freedom, the atoms, are very useful. It's quite fruitful to know about their existence.

Thermodynamics goes beyond just description. It's not kinematics. It really is dynamics, and that means we're going to need some fundamental laws to help us make sense of what's going on. In a certain sense, we can use the fundamental laws that we've already built up, the classical Newtonian laws. We will introduce some laws of thermodynamics. There are only a few of them, and they are re-formulations that help us to make sense of what's going on in thermal systems. In thermal systems and thermal physics, it's everywhere. It's of enormous practical importance. Just think about your home and insulating it in the wintertime and cooling it in the summertime. Then you go in the kitchen and you turn on the stove. You'd like the food to heat up quickly but not too quickly. You want to measure the fever of a small child. You want to predict the melting of glaciers on Greenland—many, many things in the world rely on an understanding of thermodynamics. It's a lovely field to be wrapping up this course with, because thermodynamics really does pull together all of the big ideas of classical physics. We will think about F=ma. We'll think about energy principles. We will think about the atomic hypothesis. We will think about statistical mechanics. This idea that microscopic motion can explain macroscopic thermodynamics is a lovely and powerful connection. You can talk about thermodynamics without ever talking about atoms. You can put a pot on the stove, heat it up, make measurements of the temperature as a function of time and never talk about the atoms; but we are going to think about the atoms, because although it historically took some time before the atomic hypothesis was folded in with the pure, straight observations about temperature and heat, in the end when you do put them together, it makes everything make so much more sense.

Thermodynamics begins with energy and energy flow. That's really one of the languages that helps us to understand heat and heat phenomenon. When you have a complex, real-world system, any normal object has lots of atoms in it. A pot of water might have a million billion water molecules in it. If you ask how I could understand what's going on, the reductionist viewpoint would say, all right, go all the way down. Look at the atoms. Look at the forces between them. Track them. There is an object that starts here. Now it feels a kick so now it goes that way. That might work, but, boy, would that be tough. And nobody has really ever thought about in a practical way understanding a pot of water by going down to the individual water molecules.

Instead, and this is really the big idea of thermodynamics and statistical mechanics, think about averages. Think about the behavior of typical atoms, and don't worry about the details. This is the simplification that's going on in thermodynamics. If you focus on the average, thinking about energy will definitely help because it gives us a nice, concrete, measurable handle, then we're going to be able to make sense of what's going on. Think about the insurance company that can make a living very confidently by making predictions about when people are married, when people have babies, when people are sick and how often they are sick, and when they die. It's not so hard to make these predictions on average. It's impossible to make these predictions for an individual. You could never do it. Insurance companies don't try to make a specific guess about you. Instead, they insure many, many people, and they figure everything will wash out. That's a very powerful idea, and it works very well.

Atoms are much simpler than people, so it's nice. If the insurance company was dealing with something simpler than a human being, they would have an easier job, and it becomes easier and easier as the insurance company insures more and more people. My university had a little bit of trouble a few years ago because we went with a small, local medical insurance company, and the fact that they were only ensuring the local community meant that one or two freak accidents—somebody who has a heart attack much younger than you would have expected and then costs the company a lot of money—

can have a big impact. When you're averaging over hundreds of millions, let alone millions of billions or billions, then all of the sudden the details of the individuals hardly matter at all.

Thermodynamics, the study of the understanding of temperature, heat, work, and energy in macroscopic systems is characterized by three laws, and then after the third law people looked back and realized that really there was another law that should have come first. Instead of re-numbering them, they added the *zeroth* law of thermodynamics. It's silly, but we're going to start with the zeroth law of thermodynamics. In today's lecture we'll talk about the zeroth and the first laws, and next time we'll talk about the second law. The third law is a technical and small addition. We won't really focus too much on it because the second law contains the essence of that part of the story.

The zeroth and first laws are really about work and energy and energy flow. The second and third laws are going to add a new concept, which we're going to call entropy. It's a whole new topic so we want to leave that until we need it. Today I want to pull together all of the ideas of energy and see how adding just a little bit of new language can help us to make sense of much more interesting systems than we have been looking at before.

This study has gone on for a long time. You can just imagine when you think about how practically useful it is to understand thermodynamics that people must have cared about this back thousands of years ago. You want to keep warm in the winter. You would love to be able to keep the ice from the winter into the summer. There are lots of applications of understanding thermodynamics.

In the old days, people developed a theory, and the theory went by the name of *the caloric theory*. The caloric theory was quite believable for a long time, and it has been disproved in the classical physics era. The caloric theory was discarded. It was a theory, which said what does it mean to say something is hot and something else is cold? The caloric theory says what it means is that the hot thing has some stuff in it. The caloric is a fluid, a physical, material fluid. Hot things have more of this fluid. Cold things have less of this fluid. It's nice because then when you think about pouring fluids, it's like flowing heat. A hot object heating up a cold object you could think of as the transfer of this mysterious fluid from one to the other. It's a nice idea. It meshes with intuitions about flow. It just turns out not to be correct. There are consequences of that model, which are demonstrably false.

The basis for a contemporary model came from a physicist named James Joule. We've talked about Joule before. We named the unit of energy after Joule. One joule of energy measures a certain amount of either energy or, if you prefer, work done, force times distance. "C" James Joule didn't invent this idea. Other people had thought about it and talked about it, and he was attributed most of the credit partly by historical accident and partly because he was just well situated. He did some excellent, high-quality experiments. He defended them very well. He wrote it up nicely. He was part of the physics community, and so we think about James Joule as one of the originators of this idea that when we're thinking about heat we're really thinking about the flow of energy, and we're talking about thermal energy.

We mentioned this before. When I talked about energy, I said it comes in many forms. One form is kinetic energy. That was the first form we thought about. The formula, $1/2\ mv^2$, mass times the square of velocity, tells you how much energy you have. How much energy was defined as how much work you could do, how big of a force you could apply over how long of a distance. That was one form of energy, and then we talked about gravitational potential energy and chemical potential energy. We mentioned thermal energy. It's another form of energy. Think about a book sliding across the table. It starts off with energy of motion—pure, measurable kinetic energy. It has a certain number of joules of energy. Then it grinds to a halt, and it is stops. You ask where the energy went. It did not go into gravitational potential energy. It doesn't go up, and it's not really stored in an obvious way. It's not stored like a compressed spring. You're not going to have it back if you let it sit there. There's nothing you can do to have that energy back in any obvious way. Has it disappeared from the universe? No, it's still there, and we call it thermal energy

Now that we have this atomic hypothesis, we understand exactly where it is. The friction, the force between the book and the table, started making atoms jiggle, molecules jiggle, a little bit more rapidly. If they're jiggling more rapidly, they have more kinetic energy, but it's hidden from our eye because the book as a whole and the table as a whole have no overall kinetic energy. We can't see that kinetic energy. It's random. One molecule is going one way. The other molecule is going another way, but they all have a little bit more energy each and it all adds up. That's where the thermal energy is stored. It's stored in the random kinetic energy of the molecules themselves. It's quantitative. You can count that energy. Thermal energy is measured in joules just like any other kind of energy.

When you do experiments, and you're trying to figure out whether this is a true statement or not—back in the 1700's, after Newton—people are arguing about whether it's the caloric, or physical substance, or whether it's energy that we're talking about, which is not a physical substance. It's a property of physical objects. It's a very different kind of model about what you're talking about when you're talking about heat and temperature and so on. These experiments people did were much tougher than you might think. It's easy enough to heat a pot of water, but to keep careful track of how much energy you consumed and where it went turns out to take about 200 years of steady development. Isaac Newton thought a little bit about heat and thermodynamics himself, but it was really 200 years later, in the 1800's, when people were really beginning to make a concrete physical science of all of this to realize that there are patterns and regularities. It doesn't matter what material substance you have or what circumstances you have.

If you're trying to measure, why is it so difficult? Well, think about how difficult it is to isolate a system thermally. When you buy a thermos bottle, a thermos bottle is designed to keep the coffee warm for a while, to thermally isolate the hot coffee from the cool outside world, and that works for maybe a couple of hours at which point the coffee is cooled back down again. The thermos is a relatively modern invention. It requires good seals, vacuum and high-tech materials. Back in the 1700's, it's difficult to thermally isolate objects in order to study their temperature as you add energy to them, and it's difficult to measure temperature to high accuracy. Nowadays we have very small and precise thermometers, but if you slide a book across a table, even with a pretty good thermometer you're going to have a hard time carefully measuring the change in temperature of the book and of the table. If the thermometer is like it was back in the old days, some big, bulky object with lots of fluids in it, then it begins to interfere with the experiment that you're doing. You want it to measure the thermal properties of the book, but you're mixing in the thermal properties of the thermometer itself. Of course, that's a very dangerous thing. You have to be very careful when doing physics not to be measuring the device that you're using to measure the physics. You want to measure the physics itself.

Mr. Joule was one of the first physicists to come up—systematically, carefully and quantitatively—with very clever experiments, and it is partly for this reason that we have named the unit of energy after him. Many physicists worked on this idea though. It took a lot of years, and how are you going to show that the caloric theory is wrong? Well, let's think of some very simplistic arguments against it. Number one argument: if caloric is a physical substance and you're pouring it from one thing to the other when you're transferring thermal energy, then as your coffee cools down, it should become lighter because it has less caloric in it. You could make some careful measurements, weight the coffee before and after it has cooled off, and you'll discover that there is no measurable change in its mass.

Now you have to start standing on your head. You say, well, maybe the caloric is really, really light and so we can't measure it. It's just not a massive substance. All right, supposing that I take a block and it's cool and I take a drill bit, which is also cool, and I start drilling into the solid object. It's a very, very, very rigid, solid material so the drill bit is just grinding, and grinding, and grinding and barely doing anything. You know what happens. It becomes very, very hot very rapidly. If you believe in the caloric, where is the caloric coming from? You're creating it out of nowhere. You had two cool objects, and doing physical, mechanical work—force times distance—frictional force scraping across the object at the drill bit is creating caloric out of nowhere. If you're a classical physicist, you find that difficult idea. To create a physical substance out of nothing doesn't seem as if it fits in with Newtonian ideas. Ultimately, there were many, many experiments, and this idea of joule that we're talking—not about a material substance, about energy—really became so well verified that it formed the laws of thermodynamics.

Let's talk about them. The zeroth law of thermodynamics is in "A" sense defining what thermal equilibrium means. If you have two objects, "A" and "B", and you touch them together, they might be in thermal equilibrium or they might not be. If they're not, one of them will change. It will cool off, and the other will change. It will warm up until nothing happens anymore. When nothing more happens, they're in thermal equilibrium. The zeroth law of thermodynamics says if "A" is in equilibrium with "B" and then you separate them and you check and you discover that "B" is in equilibrium with "C". All right, so we've done "A" pair of experiments. Then, the zeroth law of thermodynamics argues or concludes that "A" and C will be in equilibrium with one another guaranteed. If "A" equals "B" and "B" equals "C" then "A" equals "C". Now, you might think that's obvious, but it's not equal as a mathematical function. It equals meaning in physical equilibrium. It's a statement about the world, and it's a very practical statement. It means that if you take a thermometer and go to the factory and calibrate it, you hold it against an object that somebody has defined to be 98.6 degrees, you wait until they're in equilibrium and then you draw a little line where the fluid is. Now you've separated "A" and "B". You move the thermometer and put it in your mouth. You wait until the thermometer comes into thermal equilibrium with your tongue, and you look at the state of the thermometer and say, oh, it's in exactly the same state as it was before so the temperature of my mouth is the same as the temperature of this artificially defined definition of what 98.6 is going to mean. This allows us to use thermometers reliably and understand that they are measuring something physical and repeatable. Temperature becomes meaningful with the zeroth law of thermodynamics, and it tells us how you go about measuring it. You compare things. Your mouth is the same temperature as the 98.6 standard, and so we say your mouth is at 98.6. It's meaningful. That's what the zeroth law of thermodynamics is telling us, and notice there are some very subtle ideas in here. The thermometer can be big or it can be small. I don't care. It will be in equilibrium with my mouth and with the 98.6 standard no matter what it's made of, what materials you make it of, high-tech, low-tech, big or small. The zeroth law of thermodynamics says temperature is well-defined. It's a property of objects.

Now, if you think about atoms, you're thinking about statistical mechanics, which I kind of think of as the underpinnings of the laws of thermodynamics. What I realize is that of course temperature makes sense. Temperature is measuring the average kinetic energy of the little atoms in the system. In my body, there is an average kinetic energy, and the thermometer goes into equilibrium. What does that mean? Well, if the thermometer starts off colder, that means that the atoms in the thermometer are jiggling more slowly. What happens when you bring two solid bodies into contact and microscopically one of them has atoms that are jiggling slowly and the other has atoms that are jiggling rapidly? Well, the rapid ones will whack more frequently into the slow ones and speed them up. Of course, they'll slow down in the process so the fast ones become slower and the slow ones become faster. This continues until everybody is going at basically the same average speed, and that's equilibrium. Now we have a microscopic picture that helps us to make sense of the zeroth law of thermodynamics.

In the end, when you have equilibrium, all of the atoms will have the same average energy. Now, it doesn't mean that every atom is going at exactly the same speed. Some are going faster. Some are going slower. They keep bumping into one another all of the time, but when they reach a steady state, then you have this lovely thermal equilibrium and you have a well defined temperature. Temperature has nothing to do with the material object. It comes down to the average energy of whatever particles you happen to be made of, gases, liquids, solids, and everything is made of atoms so everything has this nicely defined temperature.

The first law of thermodynamics takes this and now adds in the story of energy flow. It's really a statement of energy conservation but in a new way. There are some words that we need to keep straight, and they are subtle words. We can easily muck them up because it's this usual story about physics where the words are defined by physicists, but they also have common English usage. I've already used some of them incorrectly because in some cases we tend to be sloppy about thermodynamics words. The three words are "temperature," "thermal energy," and "heat." We've already talked about temperature. That's what the zeroth law of thermodynamics is helping us make sense of. Temperature is one thing. It's the average kinetic energy in a microscopic model. Thermal energy is not the average. It's the total.

If you have a block and it has some total amount of energy because every atom or molecule has its own little teeny weeny kinetic energy, a little teeny weeny mass; and if you add up all of those that gives you the total thermal energy of the object. When you slide the book across the table, the original kinetic energy, however much that might be, is spreading out over many countless billions of billions of atoms. Because it is spreading out over so many atoms, each individual atom doesn't change all that much. A tiny increase added up over many millions and billions of objects can add up to a reasonably large amount of energy. That's why the book slides across the table. Its kinetic energy seems to have disappeared, and the temperature barely rose because the average didn't change very much. The grand total increased by exactly the amount that we started with.

Okay, what about "heat," the last of those three words. Heat is the trickiest of them all. If you want to use the word "heat" correctly, as a physicist does, that is to say if you want to use the word as it is defined, think of it as a verb. I heat the water—that's a good usage. You shouldn't talk about how much heat is there in the water. That's a noun, and it's treating heat as if it is a material substance. You're going back to the old caloric idea that there is something there in the water. It's awfully easy to say, and I say it myself. I won't fuss on this too much, but when you have a hot object and you have a cool object, we say that the cool object becomes heated. What do we mean by that? We mean there is a flow of thermal energy. Heat really is defined as the flow of energy, thermal energy, from one object to another object. When you talk about heating up the water, you're talking about a flow of thermal energy. The total thermal energy of the water increase when you heat it.

The first law of thermodynamics puts this together in a simple statement that says if you have an object, you can do things to it in a variety of ways, but energy must be conserved. The first law of thermodynamics says energy is conserved. Let's be very clear about this. If you have an object and it has some total thermal energy and you do something so the thermal energy changes, how much will it change? It is just conservation of energy. The thermal energy increase of an object will equal how much physical work you did. That's a transfer of energy, force times the distance, that's mechanical. That's one way of adding energy plus the other way of adding energy, which is heating. Total change in thermal energy arises from work done plus heat. Work and heat are two different ways of thinking about the transfer of energy. That's the first law of thermodynamics.

It says that if you put a cold object on the stove and the stove is hot—I could turn off the switch but the stove is still hot—the water is going to become hotter and hotter over time. Why? I am putting energy into it. If you're hotter, that means your temperature is going up. If your temperature is going up that means the average energy of molecules is going up. You still have the same amount of water, so if the average of each one goes up, then the total must be going up. We must be adding energy. Where is it coming from? How do you add energy? The old post-Newtonian way was to do work, force times distance. You could run a paddle wheel through it and make friction and do some physical, mechanical work. That would be one way of heating the water, but that's not what is happening on the stove. Instead, we are just doing this heating. We are transferring the random motion of molecules on the stove top and converting that into random motion of molecules in the pot of water.

James Joule is arguing that thermal energy can be measured. It's just all one and the same thing. It's just energy, and so heat and mechanical work are really equivalent to one another. He measured the mechanical equivalence of heat. If you do a certain amount of work, a certain force times distance, you can see how much the object heats up. You can see its change in temperature, and then you can compare that with the amount of heat flowing from some temperature difference. Now we can talk about flowing heat with the exact same measurable unit, joules. We can talk about a flow of heat in joules in these two different ways, work and heat.

In the end, the first law of thermodynamics is telling us that there is a bottom line here. We can understand temperature by thinking about equilibrium. We can think about if we want the internal energies. We can go down to the level of atoms, and what we discover is in a certain sense there's nothing new here. There's no new mysterious caloric. We don't need to hypothesize this mysterious new material substance because you can understand what's going on just by thinking about energy measured in joules in these different forms.

Thermodynamics is bookkeeping at this point. We're just keeping track. How many joules did I have to start with, and in what different ways did we transfer that energy? It's very nice. Physicists love to have simple bookkeeping tools, and energy is one of those great, useful tools. It's just a number that you add and subtract.

There is more to thermodynamics than just bookkeeping, and in the next lecture we will talk about this new concept, the entropy concept that will shift our story a little bit and teach us to think about why things happen in a slightly different way. For a moment, recognize that just the first law and the zeroth law of thermodynamics allow us to understand an awful lot of practical things, measuring temperatures, measuring properties of objects as you put them on the stove, what happens what you stir them, what happens when you squeeze them, and all of these are really one in the same fundamental idea—thermodynamics.

Lecture Twenty-Three

Heat and the Second Law of Thermodynamics

The law that entropy always increases—the second law of thermodynamics—holds, I think, the supreme position among the laws of physics. If someone points out to you that your pet theory of the universe is in disagreement with Maxwell's equations—then so much the worse for Maxwell's equations. If it is found to be contradicted by observation—well, these experimentalists do bungle things from time to time. But if your theory is found to be against the Second Law of Thermodynamics I can give you no hope; there is nothing for it but to collapse in deepest humiliation.

— Sir Arthur Eddington.

The Laws of Thermodynamics:
First Law: You can't win.
Second Law: You can't break even.
Third Law: You can't get out of the game.
— A popular (and fairly accurate) scientific joke

Scope:

Thermodynamics began as an application and extension of the basic idea of energy conservation, and the first law of thermodynamics is a mathematical statement about conservation of energy. The most significant application of thermodynamics in our lives is a *heat engine*. This is a generic term for a device or system that converts thermal energy into useful work. In order to discuss heat engines, we will need to look at the concept of *entropy*, an abstract, elegant, and powerful property of systems. Entropy can be defined in several ways, related to heat flow and temperature or to "randomness." The implications arising from entropy considerations are profound and practical, expressed colloquially as "You can't win, you can't break even, and you can't get out of the game." Entropy and the laws of thermodynamics help us understand why heat engines are limited in efficiency (and why perpetual motion machines are impossible), despite all the creativity and best efforts of engineers and inventors. These principles describe the natural tendency of isolated systems toward states of more disorder and even touch on the direction of the "arrow of time."

Outline

I. Let's begin by exploring some of the broad properties of heat engines.

A. A heat engine has a working material, a hot reservoir to take in thermal energy, and a cold reservoir to release thermal energy. The working material (for example, gas) converts thermal energy from the hot reservoir into mechanical energy. An example would be a car engine.

B. A heat engine can also extract thermal energy from a cold reservoir and exhaust it to a hot reservoir. An example of this would be an air conditioner or refrigerator.

II. By the 1800s, people were using the principles of heat engines to build steam engines, but the engines exhibited poor efficiency.

A. We can define *efficiency* as "what you get" divided by "what you pay for." For an engine, "what you get" is the amount of mechanical energy provided by the device; "what you pay for" is the amount of energy put into the device, measured as chemical potential energy.

B. With conservation of energy, we know that "what you get" can never be greater than "what you pay for." In other words, according to the first law of thermodynamics, the efficiency of a heat engine can never exceed 1.

C. In practice, if you put 1000 joules of stored chemical energy into a car in the form of gasoline, you might get 200 joules of useful work (kinetic energy) out of the system, for an efficiency of 20%. The remaining 800 joules is in the form of exhaust heat, a thermal energy increase of the cold reservoir (which in this case is the atmosphere.)

III. The second law of thermodynamics tells us that the maximum efficiency of a heat engine is generally much less than 100%.

A. This law was discovered by Sadi Carnot (1796–1832), a French engineer working with steam engines around 1800. Carnot was able to think about the fundamental principles governing steam engines, looking beyond the details of what fuel was used or what materials the engines were made of.

B. Carnot recognized the simple fact that, in an isolated system, hot objects always spontaneously cool down and cold objects always spontaneously warm up until they reach equilibrium.

C. Why does this principle never work in reverse? With two objects, why doesn't the hot one get hotter and the cold one get colder? Such a phenomenon could still conserve energy: The total thermal energy of the hot object would increase, but the total thermal energy of the cold object would decrease. The fact that this doesn't spontaneously happen *is* the second law of thermodynamics.

D. Carnot recognized that the second law had many concrete consequences.

1. The maximum efficiency of a heat engine is determined by the temperatures of the hot and cold reservoirs.
2. For a steam engine operating with fuel (boiling water) at 100° C and exhausting to room-temperature air, the maximum possible efficiency is a depressing 20%.
3. This is a fundamental law of physics, and no heat engine, no matter how well engineered, can beat this limit.

E. The statistical mechanical view, which we'll discuss in a few moments, helps us see this law even more clearly; the second law arises from how atoms move and rearrange themselves, and there is no way around it.

IV. The second law of thermodynamics can be stated in many different ways. The concept of *entropy* is a way of reframing the second law.

A. Entropy is a measurable quantity of systems, much as energy is. The statistical mechanical worldview teaches us to think about entropy microscopically as a measure of randomness. The more orderly a system is, the less entropy it has and vice versa.

B. What do we mean by *randomness*? Microscopically, the term refers to how many "states" are available in a system.

C. Adding energy is one way of increasing entropy, but the two are not the same properties.

D. The second law of thermodynamics is a counting argument, based on pure probability.

1. If you buy a new deck of cards, it is highly ordered (low entropy). When you shuffle the deck a number of times, the cards will become more random (increasing entropy), but you will never, in your lifetime, shuffle the deck back into its original ordered state.
2. With 52 cards, there is an absurdly tiny chance that the cards could be shuffled back into order, but with a million billion billion atoms in a pot of water, the chance of entropy spontaneously decreasing through natural events is 0 to any degree of approximation!

E. We can see this in relation to Carnot's statement of the second law of thermodynamics.
1. In isolated thermal systems, hotter objects get cooler and cooler objects get hotter.
2. If we're thinking about entropy, the cooler object could not become even colder, because that would be an increase in entropy. (Colder objects tend to be more orderly.)

V. Once we think about thermodynamics in this way, that is, with regard to systems and their natural evolution, we realize that some forms of energy are more *useful* than others.

A. More useful energy has less entropy. Every molecule in a car traveling down the highway is going in the same direction with the same average speed. This system is about as orderly as we can imagine, and the state is one of very low entropy for that amount of energy to be configured in. If the car crashed, the energy would be conserved, but the thermal energy would be in a much more random state.

B. This tells us why engines have a maximum efficiency. The hot gases in a car engine are in a state of random thermal energy. The engine extracts some of that energy to drive the car, but all of the energy cannot be transferred to motion of the car, because that would violate the second law of thermodynamics. Converting all the random thermal energy in the gas to kinetic energy of the car would decrease the overall entropy of the system.

VI. The second law of thermodynamics does not say that entropy never decreases. It says only that entropy of *isolated systems* never decreases.

- **A.** When you put water in the freezer, the water molecules spontaneously freeze, becoming ice. That's moving to a state of more order, or lower entropy, but it happens because the water is not isolated; freezing the water requires external energy.
- **B.** Some people have argued that human beings are low-entropy systems. "Making" a human involves increasing the order of certain chemicals to form organs; locally, then, entropy is decreasing. As a baby is formed, however, the mother requires energy from outside systems, and at the same time, she is increasing the entropy of the environment by releasing waste heat. Overall, entropy in the universe is increasing!

VII. All of this leads to a philosophical question about time: As time goes by, entropy is increasing, and we always think of time as going forward. The "arrow of time" points in only one direction.

- **A.** Newton's laws don't tell us why time should go one way and not the other. Watching a movie of billiard balls colliding on a table, we couldn't tell whether the movie was running forward or backward. Newton's laws are *time reversible*.
- **B.** But thermodynamic processes are not time reversible. If we watched a movie of an egg falling to the floor and breaking (and heating the floor slightly), we would know very well whether the movie was running forward or backward. We know that we could never use the thermal energy from the floor to repair the egg and sending it flying back upward into a person's waiting hands.
- **C.** It's possible, then, that the direction of time is associated with the second law of thermodynamics.

Essential Computer Sim:

Go to http://phet.colorado.edu and play with Friction and Reversible Reactions.

Essential Reading:

Hewitt, second half of chapter 17.

Hobson, rest of chapter 7.

Recommended Reading:

Cropper, chapters 3, 9–10.

Lightman, chapter 2.

Questions to Consider:

1. Explain how the laws of thermodynamics are, in a sense, equivalent to the joke lines quoted at the start of this lecture. (We didn't cover the third law, so just focus on the first two.)
2. Can you come up with several examples of naturally occurring processes that show how high-quality (usable) energy degrades into lower-quality (internal, thermal) energy?
3. When you clean up a room, what is happening to the entropy (randomness) of the room? Does this process violate the second law of thermodynamics?
4. Think about watching a movie of natural events, but you don't know if the movie is running forward or backward. Come up with some events where you could not tell which way the movie was running and others where it would be obvious. Now think about entropy—does it help explain the difference between these classes of events?
5. If a company claims it will produce a new super-efficient gasoline-powered engine that will put out 200 horsepower (hp) steadily for an hour on 1 gallon of gas, are you interested in buying the company's stock (or its product), or should you report the company to the Better Business Bureau as scammers? (Useful data: One gallon of gas contains 36.6 kWh of stored chemical energy, and 1 hp is 750 watts.)
6. If a company claims that it can rebuild your carburetor to make your car 90% efficient at converting gas energy into energy of motion, are you interested, or are they scamming you?

Lecture Twenty-Three—Transcript
Heat and the Second Law of Thermodynamics

Thermodynamics began as an application and an extension of the basic idea of energy and energy conservation. Thermodynamics really started off by arguing that thermal energy is just another form of energy. We don't need to hypothesize some mysterious new substance. We can just think about the flow of energy in order to understand objects warming up and cooling down, and the first law of thermodynamics is just a mathematical statement of conservation of energy. There are many, many applications of this idea, and the most significant that I can think of in our lives would be something called a heat engine. Heat engine is a generic term. It's the device that drove the industrial revolution. It's what powers our homes. It's what runs our cars. A heat engine is a device that converts thermal energy from one place, moves it, and does something with that energy. That is really the idea of a heat engine, and this is what we want to talk about today.

In order to talk about it, we're going to have to introduce a new concept, the concept of entropy, but let's just begin by thinking about the broad properties of heat engines. They're made of many, many different kinds of designs. I can think of lots of different kinds of heat engines. You always have a hot bath. There is some place that is hot in your automobile. You're burning gasoline, and you have a little hot spot. There's always a cold spot. We call it a reservoir, and this cold reservoir is the exhaust. In your automobile engine, the cold spot would probably be the outside world that is at a cool temperature. Then there will be some working material, and the idea of the heat engine would be that the working material—it might be a gas or a liquid as there are all sorts of different kinds of designs—is going to take some of the thermal energy from the hot stuff and do something. If what you do is convert that thermal energy into work, that is to say into mechanical energy—I should really use the proper word—if you convert the thermal energy of the fuel or the hot spot into mechanical energy, kinetic energy, that would be like an automobile engine taking the hot gases, driving pistons and moving the car. You're moving energy from a thermal place into kinetic energy, a different form. You could, in principle, move the energy, the thermal energy, from a hot place to a cold place, or from a cold place to a hot place. Those would also be heat engines. When you're thinking about air conditioning your house or running your refrigerator, all you are doing really is moving thermal energy around. You're taking it out of the water, turning the water into ice cubes and dumping that thermal energy into your kitchen. That is another kind of heat engine.

In order to understand heat engines, remember that in the old days, people were just building these things and they didn't really understand them. Two hundred years ago, people were designing steam engines, and there wasn't a good, rigorous physics underpinning for them. They were tinkering. There were some engineers. There were some principles that were being developed, and people began to build steam engines. By 1800, you could heat up water and then use that hot reservoir of water to drive a piston. That's the working fluid, and that pushes the wheels through some crank system. You could convert the thermal energy into energy of motion, and people were becoming more and more frustrated because they knew that they were putting in a lot of energy, they were burning a lot of coal, and yet they weren't putting out a whole lot of energy. They were not climbing big hills. They were not going very fast so there was poor efficiency. This was a lovely point in the history of science where tinkering, engineering and physics all began to work together so that there was this practical drive for the physics. The theory and the development of the laws of thermodynamics were fueled and then became immediately useful for people to understand how to apply this stuff. It's a lovely interplay, which continues to this day. I think it's part of what it means to be doing—certainly classical physics, and maybe any kind of—physics. The question that is coming to the forefront with these steam engines is efficiency. We're going to define efficiency glibly as what you have, divided by what you paid for. For an engine, what you have means how much mechanical energy the device provides you. How high up does the object go or how fast does it go? That's the useful energy. That's the numerator. The denominator is how much energy you put in, measured as chemical potential energy of the coal, or the gasoline, or whatever source of energy you started with.

Think about conservation of energy. What you have can never be greater than what you put in. What you take out can't be greater than what you put in because that would be an obvious violation of the first law of thermodynamics. The first law says the efficiency, what you have, divided by what you paid for, will always be equal to one, or more likely in a practical situation, less than one, because you can always throw away some of that energy.

Imagine if you're burning coal, and it's not really doing a whole lot. It's just heating up the atmosphere. That's not useful energy for a steam engine. Now, it might be useful energy if you're warming up your house, so the efficiency might be different depending on the purpose, but if you're thinking in particular about heat engines as real engines, then you can see that practical engines could be much, much less efficient than one or 100%. If you start off with a small amount of gasoline that has 1,000 joules of stored, chemical potential energy in it, it is well defined. You have a certain amount of gasoline. You know how much energy there is, and you know in principle how much work you could take out of it. In practice, when you burn it in a real-life automobile, you won't take anywhere near that 1,000 joules out again in the form of kinetic energy of the car. You might, if you have a really efficient car, a couple hundred joules. If you have joules of useful work and you put in 1,000 joules in the form of raw, chemical potential energy, the efficiency of that engine is 200 divided by 1,000. That's 20%, and that means that 800 joules, by conservation of energy by the first law of thermodynamics, must have been thrown away. Now, it doesn't disappear from the universe. Where does it go? It warms up the cold reservoir, which in this case is the atmosphere.

Because the cold reservoir is so big, 800 joules of energy isn't very much spread out over countless billions of atoms. They don't have to change their temperature very much and maybe even not noticeably and yet you still have conservation of energy. Conservation of energy has put a limit of 100%, and back in the early 1800's people were looking at 1% of engines at best. They were really struggling to be efficient.

There is a new law of physics, the law that we want to talk about today, in which we discovered that you can't even approach 100%. It's much, much worse than you might have thought. The second law of thermodynamics gives us a mathematical formula that tells us what the maximum efficiency is of a perfect, ideal heat engine. This was discovered originally by Sadi Carnot. He was a French engineer, and he was very interested in steam engines. He also did some good physics, and part of the brilliance of Carnot back in 1800 was his ability to look at all of these different steam engine designs and step back from the details. Never mind what fluid they use. Never mind what fuel they use. Never mind the mechanical design of the pistons. Just think about the broad principles, the laws of nature that are involved, and what he recognized was so simple that you may not even be able to imagine how this could put a quantitative limit on efficiency. What Carnot recognized was that all of the engines work in the same fundamental way. There is thermal energy in a hot reservoir, and you take that energy and do something with some of it. You are do some useful work and turn that into kinetic energy, and you also throw away some of it. You always throw some of it away into a cold bath, into the cold reservoir.

What Carnot recognized was the physical underlying principle of any design, and what always happens in nature, always, is that warm things spontaneously cool off and cold things spontaneously warm up when they're brought together but isolated from the rest of the world. It's a law of nature. It's the second law of thermodynamics that if you have an isolated system and one part of it is hot and the other part is cold, the hot part will cool while the cool part warms up until you reach equilibrium. Now, think about that for a second. It's a new law of nature. It may be obvious. It's obvious from our experience, but it's useful to recognize the fundamental nature of obvious things. Why couldn't it happen that you have two pieces of an object, maybe two buckets of water, and maybe one of them is a little cooler than the other, and the cold one just becomes colder still and the warm one just becomes warmer still? Because of their spontaneous contact with one another, why couldn't you have the cold grow colder and the hot grow hotter? It could still conserve energy if the total thermal energy of the cold one decreases and the total thermal energy of the hot one increases you could have the amount balance and preserve, the first law of thermodynamics. The fact that it doesn't happen is the second law.

Carnot recognized this, and his brilliance was that he recognized there were many concrete consequences of this. One of the consequences was that if you run a heat engine and you have a certain temperature of the hot bath, and you have a certain temperature of the cold bath, you will be limited as to what fraction of the thermal energy, stored in that hot bath, that you can let out. What is the maximum possible efficiency? Carnot used logic and mathematics and argued that the maximum depends only on the temperatures. That's the best efficiency you could ever come up with. We're talking about the perfect design, no friction, no internal problems, no poor design, and even in this ideal case the maximum possible efficiency will be limited. It's a simple formula. The efficiency basically depends on the ratio of the cold temperature to the hot temperature. If you start off with boiling water, that's a steam engine and the boiling water is at 100 degrees Celsius and its running some sort of cycle, you're pushing on pistons because they're hot now and they're expanding gases and they're pushing on iron rods, and then you have to cool the gas back down again in order to repeat the cycle. You have to always have a cycle if you want the engine to keep running, and that's part of Carnot's insight.

If you look at the temperatures involved and you look at Carnot's formula—which I haven't derived for you but I'm just stating—the maximum possible efficiency is very depressing. For a steam 25%, for every useful joule, you throw away another three. That seems very painful, and at first the engineers working on steam engines said, oh, I realize that right now we're only at 2%. But we're not shooting for 25. We're shooting for 100, or let's be realistic, 99.99. They weren't being realistic because they didn't recognize that Carnot's rule is not an approximation. It's another one of those fundamental laws of classical physics. In the end, if you think about the world being made of atoms, and we talked a little bit about this, the statistical, mechanical view of this thermodynamics story makes the story even more rigorous. It's really a law of nature that arises from counting how atoms rearrange themselves and how they're moving, and there's no way around it.

This is not a law of physics that you can struggle from underneath any more than conservation of energy. It's a statement about how the world works, and so not only should you not buy a free-energy machine—that doesn't take any fuel, but produces some sort of output—but you shouldn't even invest in a machine that has 90% efficiency by burning something, because if it's burning something, it's a heat engine. If it's a heat engine, you can't possibly have 90% efficiency. Mr. Carnot's formula tells you that to do so, your hot temperature would have to be so hot that ordinary objects would vaporize. You couldn't make it in any practical sense. Yes, you can increase the efficiency of engines, but to do so you have to make the temperature difference much, much bigger than you normally would in ordinary working circumstances.

The second law of thermodynamics can be thought of and stated in many different ways. It's one of the pleasures of studying thermodynamics as a career, to begin to realize how all of these different ways of thinking about the second law all hook together. Some of them seem intimately related, and some of them seem really different. The one I'm about to tell you is a new concept it's a way of reframing the same second law of thermodynamics. It's the concept of entropy. Entropy is a number. You can measure it. If you have a pot of water, it has entropy, and you can measure the energy. Now, energy, of course, is also a fairly abstract concept. How would you measure the energy? You would need to really have some deeper understanding of what's going on and what energy means. Energy tells you how much work you can do, so measuring energy is a tricky business. Measuring entropy is an even trickier business. Now, what is entropy measuring? In the early days, in the post-Carnot days, people were beginning to recognize that there is this number, it is associated with systems, and it changes. As you do things, you can increase the entropy, or you might decrease the entropy. People came up with thermodynamics rules for measuring how much the entropy changed over time.

Ultimately, the statistical/mechanical worldview taught us a way of thinking about entropy microscopically. It's a measure of randomness. It's a measure of order. The more orderly something is, the less random it is. The less entropy it has, and the more random something is, the more entropy it has. It's a quantitative measure of randomness. What do I mean by randomness? Well, microscopically I mean counting how many different states are available. States would be different configurations. One way that you could give a pot of water more available states would be to warm it up, because now the molecules are jiggling around faster, so they have more opportunities in their life. They can go a little bit faster or slower. There are more options for them, and of course you think about them jiggling around more and so this idea of randomness is nice and deeply intuitive. Adding energy is one way of increasing the entropy, but it's not the only way. Energy is not the same as entropy.

If you think of little molecules flying around in a chamber, if the chamber is bigger and those molecules have more space to run around in, they have more options available to them. It's more random for them to be spread out over this bigger space. There are alternative ways of thinking about the microscopic details of entropy.

The second law of thermodynamics is just a statement of probability. It's a counting law of nature, and let me give you a concrete example so that we can think about it. Suppose I have a deck of cards. You buy it, and it comes highly ordered. When you buy it from the store, you take off the plastic, it is ace through king of one suit, ace through king of the next suite, and that way for all four suits. It is ordered. You can predict by looking at one card what the next card will be. That's part of what you mean by an orderly system. If you were to now shuffle that deck, or just drop them on the floor and pick them back up again, if you do something with the deck, then they will become more random. That's what spontaneously happens, and that's the second law of thermodynamics. The second law of thermodynamics says if you just let systems evolve and if they're isolated, then as time goes by either the entropy will stay the same or it will grow bigger. Things become more random over time, but they won't spontaneously decrease in entropy. If you take that deck of cards and you don't shuffle it at all, there is no change in the entropy. You can sit there and hold it, and you can move the deck around without mucking up the order. Entropy doesn't have to increase, but as soon as you shuffle it, with the first shuffle the entropy increases somewhat. Now it is much more difficult to predict what the next card will be from the previous one. If you look carefully, after one shuffle, you will still see some order, but it will be a little pattern such as two, three, four and then eight, nine, and 10. After a couple of shuffles, it's really random, and from then on shuffling more and more and more won't increase the randomness. It's already fully shuffled, but what won't happen, this is the second law of thermodynamics, what will never happen in your lifetime is that you take a random deck, shuffle it and you look at it and go, huh, ace through king of spades, ace through king of clubs, in order. It's not going to happen. Why won't it happen? Well, because its probability is insanely unlikely to happen. You can count how many states are available with 52 distinct cards. How many different orders are there? That's a fun little statistical calculation, and you can write it down. There are a lot of zeroes in the answer to that question, and the probability that a random re-ordering will put it into this ordered state is basically one out of all the many possibilities. Only one is the ordered state, and so the probability of that happening is one out of this astronomically huge number. It's so improbable as to be practically impossible. Now, that's the case with 52 cards.

If we go to one extreme and we only had a couple of cards such as simply ace, two, three and four, and you shuffle them it's going to be more random, maybe ace, four, three, two and then two, four, three, ace. If you shuffle enough times, sooner or later you might be back to ace, two, three, four because there are only four so the probability is small but it's certainly not zero. With 52, you could shuffle once per second for the rest of the lifetime of the universe, and it's not likely to happen.

Now, imagine going not from four to 52 but from four to 52 to a million billion trillion, because that's a pot of water. The thermal motion, the random motion of the molecules, is really shuffling the deck. It's shuffling the ordering of the molecules in the water, and now instead of 52 you have a million billion trillion and the number of states available, just as the number of states, is much more than 52. There are not 52 configurations of cards. There are billions of billions, and the number of configurations of the molecules is an insanely large number. If you tried to write it down, you could start writing zeroes and walk all the way to the edge of the universe, and you' would not be done yet. What is the probability that those will go into a more ordered state? The answer is zero. That is the second law of thermodynamics.

If you think about this, you can realize that this really is connected with the way we were thinking about the second law of thermodynamics, which is Mr. Carnot's argument. The first statement that I made was that thermal systems, when they're isolated, always tend to go from hotter things to colder and from colder things to hotter. I can understand that if I'm thinking about entropy, because spontaneously, if the colder thing kept growing colder and colder, it would be more and more orderly. It would go to perfect order if this kept on happening. Now, it is true that the hot thing would be more and more random, and so you might ask, well, could the increase in entropy of the hot thing more than balance the decrease in entropy of the small thing? It becomes a technical counting story. You have to think about how many states are available. It depends on the temperature, and what you'll discover is that Carnot is absolutely right. It's pure mathematics. It's derivable.

Once you think about thermodynamics in this way, once you think about the second law as a statement about systems and their natural evolution, you realize that there is energy and there is energy. Some energy is more useful than other forms, and what do I mean by more useful? Now we can be rigorous. More useful energy means less entropy. If you have some material that has energy stored and that energy is very orderly, for instance in an automobile that is traveling down the highway, every molecule is going in exactly the same direction with the same average speed. It's the speed of the car. That's about as orderly as you can have it, and that's a very low entropy state for that amount of energy to be configured in.

What's going to happen over time? Well, spontaneously that car might crash. If it crashed, then that energy would be conserved, but it would now be random thermal energy. The car is stopped, and now the molecules jiggle faster. You have taken that same amount of energy, and it is spontaneously turned into a more random distribution. That's the way things go. What never happens is nature, it's the second law of thermodynamics, is that a car crash spontaneously undoes itself, the metal reforms into a more orderly care, and then all of the thermal energy from the crash is converted to kinetic energy of the car. It won't happen. That's the second law of thermodynamics.

Once you understand this, you recognize why there is a maximum efficiency for an engine. Imagine that you have the hot gases in your automobile engine, and so you have some energy in there. Every molecule jitters around fast. There is plenty of kinetic energy if only you could extract that energy and put it into your car instead of it being random in the molecules that would be great. That would be how you would drive your car, and you can do that. That's what the car engine is doing, but you can't take all of the energy and put it into the motion of your car because that would violate the second law of thermodynamics. Think about it. If you started with purely random energy, it has high entropy, and if you could somehow convert it all into kinetic energy, then you have just radically decreased the entropy of your automobile. You started off with high entropy and spontaneously—through some workings of a magical engine—convert it into a low entropy system. It's never going to happen. There is no such magical engine.

The second law of thermodynamics says you have that energy, and you can't convert it all into mechanical energy. You have to throw some of it away, increasing the entropy of the cold bath, and that will increase in entropy enough to cancel the small decrease that you have by also ordering the molecules in the car. There is a balancing act, and the second law of thermodynamics tells you quantitatively how much the best balance is that you can have. That is Mr. Carnot's original formula where he wasn't thinking about randomness. He was just looking at the steam engines themselves.

There is a very common misconception about the second law of thermodynamics. Many people misread it, and they miss that caveat for an isolated system. If you don't read that, then you say, oh, the second law of thermodynamics says that entropy can never decrease. If you thought that is what the second law of thermodynamics says, then you would become very confused because if I handed you a deck of cards that was random, and you sit down for a few minutes, you can order them. You can decrease the entropy of that deck of cards, and you would say, huh, is that a violation of the second law of thermodynamics? The answer is no. It took some work. The cards are not an isolated system. They are not spontaneously re-ordering. You need to do some work on them. You can locally decrease entropy. Put the water in the freezer, and they will spontaneously, by themselves, without your doing anything further, become colder and colder and then would freeze. As they become colder, they're becoming less entropic, less random, and when they freeze, the entropy is going down even more. Now they are very ordered. There is a crystal. If you know where one molecule is, you know right where its neighbor is. It's a very orderly state.

Did that happen spontaneously? Well, yes and no. Inside of the freezer it happened spontaneously, but look at the system that it's part of. You have to plug the freezer into the wall for this to happen. If you don't plug the freezer in, it's not going to just happen by itself. You need some external energy.

Some people have argued that human beings are low entropy states because we're very ordered. We have organs that are separated, and if you think about the chemicals that made up your body, if those chemicals were all randomly separated, they would have much higher entropy. That is correct. When your mom made you, when her body created you, she was taking chemicals and ordering them into the shape and form of a human body. Entropy was decreasing in your location, but mom was eating. She was metabolizing. She was consuming energy from an outside system, and she was throwing away a lot of waste heat. In fact, she was warming up the environment that she lived in, vastly increasing the entropy of her surroundings—orders of magnitude more than the decrease that you represented. Yes, you can decrease entropy here as long as you increase it somewhere else, so that the sum total is always going up.

Now, this has led some people, philosophers largely, to think about time. Time is connected to the story because as time goes by, entropy is increasing, and people have often puzzled about why. How do we think about time always moving forward? The arrow of time points in one direction. The future is very distinct from the past. If you look Newton's laws, you don't have any clue about why time should go one way and not the other. Think about watching a movie of billiards, perfect Newtonian motion. They come in. They interact. They go back out again—conservation of energy, conservation of momentum. You can't tell if they're running forwards in time or if you run the movie backwards and they're going backwards. Both look perfectly reasonable. Newton's laws don't seem to tell us which way time should go, but if you watch something more complicated, something where thermodynamics is involved—as if you drop an egg, it hits the ground and smashes—you know very well that if you run that movie backwards that you're looking at nonsense. Eggs don't spontaneously start in this random, mushed-up state. It takes energy from the floor, cooling the floor a little bit to take that energy to rebuild chemical bonds to make an eggshell. Oh, and let's use a little bit of that going into the form of kinetic energy to make it fly up into your hand. It won't happen. The arrow of time points in the direction of increasing entropy—just a fun, philosophical point to think about when you're thinking about this entropy story.

In the end, we have a new law of thermodynamics. It's important. You can't understand heat engines if you only think about energy conservation. You have to think about entropy and randomness, or the second law of thermodynamics, in order to fully understand the complexity, reality and ordinary behavior of any thermodynamic system.

Lecture Twenty-Four
The Grand Picture of Classical Physics

Poets say science takes away from the beauty of the stars— mere globs of gas atoms. Nothing is "mere." I too can see the stars on a desert night, and feel them. But do I see less or more? The vastness of the heavens stretches my imagination—stuck on this carousel my little eye can catch one-million-year-old light. A vast pattern—of which I am a part... What is the pattern or the meaning or the why? It does not do harm to the mystery to know a little more about it. For far more marvelous is the truth than any artists of the past imagined it. Why do the poets of the present not speak of it? What men are poets who can speak of Jupiter if he were a man, but if he is an immense spinning sphere of methane and ammonia must be silent?

—Richard Feynman, footnote in *The Feynman Lectures on Physics*

Before I came here I was confused about this subject. Having listened to your lecture I am still confused. But on a higher level.

—Fermi

Scope:

Classical physics is defined in part, historically and, in part, by a philosophical mindset: The world is ordered, and there is a limited set of fundamental ideas that explain and predict all natural phenomena. The world is made of matter and energy, existing in space and time, with measurable properties and behaviors. These core ideas can be quantified via a small, consistent set of assumptions and mathematical relations, with enormous practical and predictive power. The specific ideas we have discussed form just part of what we mean by classical physics: force and acceleration as "cause and effect"; energy flow as an alternative tool for thinking about natural processes; gravity, electricity, and magnetism as fundamental forces; and matter made of atoms, with optics and thermodynamics as natural consequences. The universe in this framing is deterministic and "clock-like," with complex behavior understood by a reductionist approach to first principles. This approach to scientific truth is still widely used by scientists and others working in many fields, but it is not the "end of understanding," nor ultimate truth.

The developments of nuclear physics and radioactivity led to a totally new kind of mechanics, *quantum mechanics*, which approaches the world quite differently, with different assumptions about the "rules of the game," as well as the philosophy behind the game. The ideas of relativity challenge Newton's belief in a fixed, external space-time frame in which physics "occurs." But all of these developments remain consistent with, or connect tightly to, classical physics, which will always remain as one of our grand intellectual achievements.

Outline

I. Classical physics was firmly established by Isaac Newton and has undergone continuous development ever since.

- **A.** We might define classical physics by its dates, starting in 1687 with the publication of the *Principia* and working our way up to about 1900. This definition isn't entirely satisfactory because classical physics continues to evolve to this day.
- **B.** It's probably more productive to define classical physics in terms of its topics: mechanics of particles, forces of nature (gravity, electricity, magnetism), optics, thermodynamics, and so on. Classical physics also encompasses other topics, such as fluid flow or acoustics, that we haven't talked about in this course.
  - **1.** These topics are often fundamentally about the world we live in—they apply to cars, bicycles, rockets, sports, architecture, and many other aspects of our lives.
  - **2.** But classical physics also goes beyond everyday experience; with this study, we can understand particles down to a size of a billionth of a meter and up to the distance scales of our galaxy and beyond.
- **C.** We might also define classical physics in terms of its applications. This discipline is still studied by all scientists, as well as architects, engineers, and others.
- **D.** To paraphrase Newton, we could say that classical physics is the giant on whose shoulders science stands today.

II. Classical physics is also, in part, a way of thinking about science, a scientific-philosophical vantage point.

A. A classical worldview is traditional and empirical: it says the world is real and exists independent of human beings; the goal of classical physics is to learn about this real world through experimentation and the development of coherent, unified theories.

B. A classical worldview is often reductionist, operating from the perspective that complex natural behavior can be described, explained, understood, and predicted by analyzing simpler components.

C. Related to the ideas that the world is real and reducible is the classical idea that the world is deterministic; that is, we can use our understanding of the world to make qualitative and quantitative predictions. In this view, our universe is similar to a giant clockwork.

D. Ultimately, classical physics postulates a small, cohesive set of underlying ideas. This set includes the kinematical ideas of position, velocity, and acceleration and the dynamical ideas of inertia, mass, force, momentum, and energy. It also includes some of the laws of nature that we discussed: Newton's laws, conservation laws (related to energy, momentum, angular momentum, and charge), and Maxwell's equations.

E. Finally, classical physics relies on the scientific method of investigation: observing the world, forming and testing hypotheses, and asking further questions.

III. Physics today has moved in new directions, to the realm of modern physics.

A. Albert Einstein (1879–1955) is the "hero" of modern physics, although he was, in much of his work, a classical physicist too.

1. Einstein's theory of special relativity built on Galilean relativity, the idea that the *laws* of nature are invariant, independent of the reference frame of the observer.

2. Combining this idea with Maxwell's equations, Einstein realized a radical truth: The speed of light is itself a law of nature, independent of the observer.

3. Einstein saw that neither Galileo nor Newton was completely wrong, but our intuitive understanding of space and time—the Newtonian idea that space and time were fixed and universal and independent of the observer—required revision. For example, relativity requires us to redefine momentum and kinetic energy, for objects traveling at speeds close to the speed of light.
4. In the same way, Einstein's theory of general relativity changed the way we think about gravity. It re-imagines space and time in a geometrical sense and gives us a new idea about what gravity is and where it comes from.

B. Beginning around 1900, the field of quantum physics began to delve deeper into the atomic hypothesis of classical physics.

1. When physicists tried to understand the structure of atoms themselves, it became apparent that Newton's laws were insufficient to deal with distance scales of billionths of a meter or smaller.
2. The theory of quantum mechanics was not just a "fix" of Newton's laws; it altered the fundamental premises of classical physics, including determinism.
3. Classical physics assumes that if we understand enough about the world, we can make predictions about natural phenomena, such as the weather. With quantum mechanics, we find that many measurements, such as the time required for a particle to decay, are fundamentally uncertain and cannot be predicted.

C. Do these new discoveries completely unseat classical physics? Absolutely not. Classical physics describes the world we live in as accurately today as it did in the 1600s when Newton was first making sense of it. Modern physics *builds* on the underlying ideas of classical physics, expanding their bounds of applicability.

III. Classical physics has been one of the most fruitful, productive, and powerful intellectual endeavors in the history of civilization.

A. We study classical physics, not because we're interested in history, but because it still influences much of contemporary science and engineering.

B. We can also use classical physics as a tool to understand the science behind political issues involving energy or the environment.

C. As we close this course, I hope you will keep these ideas in mind and investigate them further in your everyday experiences.

Recommended Reading:

The world is your oyster.

Questions to Consider:

1. In what ways does physics connect to your personal life? Is your interest in physics intellectual, academic, practical, or some combination of these?
2. Modern physics has changed the philosophical outlook of scientists on many levels and challenges ideas as fundamental as the absolute nature of space, or the nature of atoms, or even Newton's laws. (For example, $F = ma$, force = mass × acceleration, is not useful or even completely correct, if you consider electrons inside an atom or objects traveling near the speed of light.) Given that, what is the value in studying classical physics?
3. Do physicists still "do" any classical physics? (Who else uses classical physics?)
4. What steps do you need to take to continue to satisfy your interest and curiosity in physics and science?

Lecture Twenty-Four—Transcript
The Grand Picture of Classical Physics

Classical physics, the topic we've been studying throughout this course, was established and introduced by Isaac Newton in 1687 with the publication of the *Principia*, and it's been undergoing pretty continuous development and evolution ever since. I think it's very productive and useful to step back at this point and revisit the question of what classical physics is, and how we define and understand what it is that we've been studying. One of the ways that you might choose to define it is that it is the physics that was done starting with Isaac Newton and working your way up to 1900. I don't think that's the most productive way of thinking about it, because classical physics continues to evolve today. There are people whose career is studying and thinking about classical physics.

In 1900, new ideas were introduced, but it certainly wasn't the end, or even the change of classical physics. I think it is probably more useful to try to define it in terms of the topics, the things that classical physics thinks about.

Classical physics is about mechanics. Mechanics is the study of motion, force and energy. It is trying to understand objects, material objects, particles, how they move and why they move that way. As we saw, it goes beyond just particles. We can study forces of nature, gravity, electricity, and magnetism, and we can think about optics, waves, heat and atoms. All of these are individual topics, and that pretty much spans the set of topics, by and large, in classical physics.

You could argue that there are other ideas out there, classical ideas, which we haven't specifically talked about. What about fluid flow? Is that classical physics? Sure, fluid flow is something that we didn't specifically address, in part because we don't need to. If you want to understand how fluids work, you have all of the building blocks. It's really about energy, force and particles moving so even though there have been new specific topics, the broad, underpinning ideas, once established, formed this framework that we use to understand a wide set of questions about the world. That really is what classical physics is to me. It's the physics of the world we live in. You look around you and try to understand. You ask questions about anything that is measurable in the world we live in. You look at cars, or bicycles, or rocket ships. You look at sports. You think about architecture, and these are the everyday experiences of the world that we live in that classical physics is helping to make sense of, to understand. Even that as a definition, though, is too limited.

Classical physics goes beyond ordinary human experience. We can understand planet Earth, the entire planet, the solar system, the galaxy, the Milky Way that we live in, and even clusters of galaxies all using classical physics. It goes far, far beyond ordinary human life experiences, and it goes in the other direction, too. We can understand tiny particles. Look at the smallest grain of dust or sand, and we can describe and understand its motion and behavior by using classical physics. Take out the microscope and you can start looking smaller and smaller. You can go down to distant scales of bacteria or millionths of an inch, even smaller still. Again, classical physics still works just fine. Newton's laws, energy, conservation of energy and moment—these principles determine the properties of the world that we live in even down at these distant scales very far from our everyday experiences.

You might think about the applications. Is that what determines classical physics? Classical physics is still taught and learned not just by physicists, but by all scientists, biologists, chemists, geophysicists, astrophysicists, and even people who go into fields that really don't look directly like scientific investigations. I'm thinking of doctors, archeologists, and architects. All of these people are using classical physics and applying it to a variety of experiences at many, many different scales, and it's enormously productive. It's the way we think about the world that we live in. Isaac Newton commented that, if he has seen further, it's because he stands on the shoulders of giants, and maybe what I'm arguing is that classical physics is the giant on whose shoulder we stand today, as we move into new realms of study, into modern physics, or contemporary biology, or any of a number of modern disciplines. We are still using classical physics. We're thinking about it. It's a vantage point from which we can continue to explore the world that we live in. Maybe that's another way of thinking about what classical physics is. Rather than looking at what it is we study, maybe we should think about how we study these things because that's part of, a deep part of, what we mean when we talk about classical physics. It's really a philosophical approach. It's a scientific philosophy for studying things. If you're a classical physicist, you say the world is out there.

It's real. Human beings are learning about the world, but the world we are learning about exists independent of us. That's a philosophical idea, and you can challenge it if you wish. There are people who make very compelling arguments about this topic, but it is definitely part of what we mean when we're thinking as classical physicists. The Earth goes around the sun. It's a big, solid object, and it's out there even if you're not looking at it, even if in some scenario there were no human beings around. The Earth would still be there, and it would still go around the sun.

Now, of course, the language that we use to describe orbits, and the formulas, and Newton's laws is all definitely about human beings and our description of this reality. Classical physics is really arguing that the real world is out there, and that's the goal of classical physics to learn something about that real world. We do it with experiments, and we do it with development of theoretical frames. We're trying to come up with a coherent, unified understanding of as much of this external reality as we can.

Classical physics goes beyond just accepting the reality of planet Earth. It's also a belief that complicated things can be reduced to the simpler components. I think that's a big part. It's not an absolute, necessary part of classical physics, but it's deeply ingrained in many of the ideas and approaches of classical physics. It's a reductionist idea that if you look at something really complicated, the sun, for instance, with its enormous number of constituents it's very hot. There is a lot going on in there. You could look at a human development such as a nuclear reactor or a CD player, and again this is a very complicated object with many parts that have complex function and behavior. You might despair and say it is just too complicated. It is essentially magic, and I give up. The philosophical approach of classical physics is to say, oh, we can understand this, and we can understand it by breaking it down. What is it made of? What are the fundamental underlying ideas, and how do they fit back together again so that we can understand this complicated thing?

If you look at a nuclear reactor, you might say, oh, that's something totally new. We haven't talked about that in this course so that must be beyond the reach of classical physics, but it's not. It's very much a classical device. It's just straight thermodynamics. Instead of burning coal, you have some hot uranium, but other than that little detail, which really is in many respects just a detail, it's a heat engine. You have a hot spot. You have a cooling pond. You have working fluid, which is probably just steam in a nuclear power plant, and you can make sense of all of the complicated procedures, outputs and behaviors just by understanding the classical thermodynamics that we've been talking about, studying, and learning about in this course.

There is another aspect of classical physics, another philosophical idea that the world is deterministic. I'm going to argue that it is related to, but a little bit different from, believing that the world is real, and believing that the world may be reduced to simpler ideas. Once you've done this reduction, it's the idea that we can make predictions, quantitative and qualitative. We have an understanding of the world if we can make predictions about what is going to happen in the future. It's as if the universe, as a whole, is some giant clockwork. That's part of what it means to think about classical physics. Now, I don't mean it literally. The Earth is not ratcheted and connected to the sun in some mechanical way, but in this belief that the world is deterministic, there is sort of a clockwork nature to our behavior. I can predict where the Earth is going to be in six months, or even 60 years, or 6,000 years. I can predict eclipses. I can predict just about anything that you want to know about the planets, and the solar system, and a huge variety, of course, of other things because classical physics says the world is deterministic. It works backwards too. I can tell you where the planet Earth was and where the planets were long before there were any human beings, and certainly long before there was any Isaac Newton, by subscribing to the ideas of classical physics.

Ultimately, classical physics postulates a small and very cohesive set of underlying ideas. There is this tapestry of ideas that we've constructed, and we've really looked at the picture that was formed. We started with the fundamental ideas of position and time. These are underpinning ideas. Isaac Newton made it fairly concrete. He talked about the operational definitions that we need to talk about more complex ideas—such as velocity and acceleration—but they really boil down to understanding just space and time. Then we talked about some fundamental dynamical ideas—dynamics being the explanation of motion.

We talked about inertia as a concept and mass, which is a numerical or quantitative experimental measure of inertia. We talked about force. It's a primitive concept—primitive meaning it underpins our theory of the world—and once you accept this idea of force, then Newton's laws follow from experiment and we're off and running. We understand the world based, on these basic ideas and other ideas that we introduced, such as momentum and energy, are defined, in a certain sense, in terms of these underlying ideas. Momentum is mass times velocity. Kinetic energy is half mass times velocity squared. We've already talked about mass and velocity. Now we're just building up these more complex and rich classical theories of the world.

We introduced some laws of nature. We talked about not just Newton's laws, but conservation laws, conservation of momentum and conservation of energy. We didn't talk too much about conservation of angular momentum. We talked about it briefly, and it's the idea that a spinning bicycle wheel—which is standing on an upside-down bike, so the bike isn't moving and it doesn't have momentum—the wheel has momentum. It's an angular or rotational momentum. That's another one of those fundamental ideas that helps us to describe and understand the world around us.

Conservation of electric charge—one of the last conservation laws that we've talked about in this course—basically putting those together along with Maxwell's equations, which describe electricity and magnetism, we really have a description of the world. That's our classical world view along with the philosophical approach and the scientific method. There were a few other ideas such as the idea of atoms and the consequences of there being atoms, thermodynamics and statistical mechanics. That pretty much outlines classical physics, and with these ideas we understand a great deal about the world we live in and beyond in all directions.

Classical physics wasn't the end of the story in a lot of respects. First of all, there are people today who continue to study these ideas and develop them further. There are now new branches of classical physics. Almost any good high school science fair project likely has an experiment where a young person is curious about the world they live in. They're trying to investigate something about the world, and they're probably using the methods of, and the ideas of, classical physics and are extending them. Sometimes people discover very, very new things, new applications and new deep ideas, all built on this framework of classical physics.

It's a nice idea that when you're studying the world, you are really asking questions and being curious, and then in the process of using classical physics, not only are you answering these questions, but also you're posing more. Any good science fair project asks more questions than it answers, and that's one of the ideas behind classical physics, that we retain to this day. It's part of how we think about science and the world that we live in. We're constantly learning, and we just have this useful framework to help us, but there is always more. There are always new ideas even within the realm of classical physics.

We have moved in some respects in the 1900's, in the 20th, and now the 21st century, into a realm of modern physics, which in some respects is just an extension of classical physics. In some respects, it's a shift in the way you think about the world. I'd like to spend some time in this last lecture looking a little bit at some of the key developments of modern physics—just an overview so that we can compare and contrast the differences between modern physics and classical physics. It might help us to understand the usefulness, the power of classical physics, as it still continues today by seeing what is different about modern physics.

For instance, Albert Einstein is really the hero of modern physics, and Albert Einstein was in many respects throughout his life very much a classical physicist. He looked at these old ideas, respected them, and in many cases believed them so strongly that he was able to build them farther. One of his earliest examples was the theory of special relativity. Now, you might think that's a whole new branch of physics that has nothing to do with classical physics, but on the contrary, it was really developed by Galileo. Galileo invented relativity, or at least he is one of the inventors of relativity. We talked about this. Galileo argued that there are laws of nature out there, and we are learning about them. Those laws of nature are, to a certain extent, independent of humans and who is observing them. This was a very big and important idea in classical physics. It says that I can observe an experiment and you, an independent observer, can be moving past me with a constant speed and straight line. We are in different reference frames. We make measurements. We disagree on many things. Speed is relative. Velocity is relative. Position is relative to the observer, but the laws of physics are invariant. They are the same. We both agree that $F=ma$.

Albert Einstein thought about this, and he believed it. He understood that this was a statement, a profound statement, about the nature of the world we live in. He didn't change that idea. He also looked at Maxwell's Equation and said these two, these classical ideas are very deep and very profound, and they seem to be matching experiments so well that they seem to be telling us something true about the world that we live in.

Now, if you combine Maxwell's equations with Galileo's principle of relativity, you discover something very interesting and very radical at that time. Maxwell's equations tell us that the speed of light is a constant independent of the observer. The speed of light is itself a law of nature. You might just accept that as a fact, but you have to think about it. It's weird. I just said speed is relative to the observer. Imagine that you are running away from me at very, very high speeds. You are running away from me at three-quarters the speed of light, and I shine a flashlight beam in your direction. I measure the speed of that light beam to be the speed of light, 300 million meters per second. You are running away from the flashlight, and you turn around and you do some experiments to measure the speed of light of that beam. What do you have? If life were simple and classical, then you would have reduced speed. If you were running away from a thrown ball and you measure its speed, it seems to be approaching you more slowly. You measure its speed to be reduced, but not light. Light is different. It's a law of nature, which Einstein recognized, that the speed of light is the same to all observers. Now, that's difficult to make sense of, and ultimately, in order to make sense of it, Einstein realized it's not relativity that's wrong. It's not Maxwell's equations that are wrong. It's our kind of intuitive understanding of space and time that needs to be revisited.

Newton believed that space and time were rigid, fixed, universal and completely independent of the observer. It was a natural belief. It forms a core idea in classical physics, and that was what Albert Einstein realized needed to be tightened up. We need to make operational definitions obeying the philosophy of classical physics in practice by trying to define precisely in the laboratory what we mean by time and time intervals. Once you do that, you realize that certain aspects of Newton's laws have to be fixed. We have to rethink our definition of time and space, which we barely thought about in this course because we just accepted them as underpinning core principles. Modern physics simply questions the definition of those underpinnings, and so it doesn't eliminate Newton's laws. They're still quite correct. It's just asking questions about where they come from and what the core of classical physics is itself.

If you think about relativity, it changes some of your ideas about space and time. It requires that you redefine momentum. It's not just "mv" anymore. You have to fix up the formula. Now, does that make Newton wrong? Not at all—when you fix up the formula, you still have effectively mass times velocity. You still absolutely have conservation of momentum. It's just that when objects are going at this extreme speed, near the speed of light, you need to make subtle changes in the way you think about what momentum means and what kinetic energy means. It is very important if you're worrying about objects that are traveling nearly 300 million meters a second, which has essentially no bearing on anything in our ordinary lives. It's nice to know about. It's a shift in your philosophical underpinnings.

Einstein's theory of general relativity did a similar thing. Remember, Isaac Newton has written down the fundamental law of gravity, and we've talked about it and understood how useful it is. Einstein didn't throw that away. He again asked where it comes from. How do I make sense of gravity itself? In classical physics, we just sort of shrug our shoulders and say, it is what it is. It's the force that attracts the Earth to the sun. Once we accept that, then we're off and running. We describe the planets, the tides, the eclipses, and on and on, everything you can think of, even rockets, to Mars. Albert Einstein says, yes, that's all great. I'm not going to throw that away, but what I would like to know is why. Why is there gravity? It's a delicious question.

In the old days, it might have been a philosophical question, and Albert Einstein turned it into a measurable question. The general theory of relativity now re-imagines space and time in a geometrical sense and lets us have this new idea of what gravity really is and where it comes from. It's fun to learn about, and I encourage you after having studied classical physics, to study some modern physics and learn about the underpinnings. One of the directions in which you might go if you push on in classical physics is to say, okay, we started classical physics at its core with atoms. We postulated that they exist. They are real. They are out there in the world, and once we believe in them, we can understand the colors of objects, the textures of objects, the behavior of objects, thermodynamics and heat engines, all of that. None of that goes away when you ask the question what is an atom made of? It's just a delving deeper. You're asking, okay, what is that atom? How do we understand it? When we investigate that, it is the field of quantum physics, which began around 1900. That is why I set 1900 as the shifting point; it was when we began to ask some new and different question. Quantum mechanics asks about the electrons inside of the atom, and here again we discover that Newton's laws are insufficient. They are inadequate to go down to distant scales of billionths of a meter or smaller. When you start investigating the sub-atomic structure, you discover to your surprise that not only do Newton's laws need to be fixed, and those are the laws of quantum mechanics, so Newton's laws are classical mechanics, but you reformulate the laws of physics and also begin to ask questions about the assumptions that we have been making, such as determinism.

In classical physics, it is taken as essentially a philosophical article of faith that if you know enough about the world today, you can make a quantitative prediction about the world tomorrow. Now, classical physics recognizes that sometimes it may be complicated, as with weather, nevertheless it is a belief that we can predict the weather. We do a pretty good job certainly for tomorrow and even two or three days down the pike, which is better than we used to do. You can easily imagine that 20 years from now, we will have even better long-range forecasts, but if you're looking at a sub-atomic particle, such as a radioactive nucleus, and we understand that radioactive nuclei decay, and you ask when—when will it decay and what is going to happen in the future—it's interesting. Quantum mechanics says we don't know, and it's not that it's just difficult to calculate or is complicated. It's that nature itself doesn't know. It's not determined today when precisely that nucleus will decay. It's a wild idea, and it turns out that we can describe nuclear decay very, very accurately on average. I can tell you if I have a million of them how many will be left tomorrow and a million years from now, but I can't tell you specifically which ones. It's like the insurance company that can tell you very, very accurately what human life spans are, when people are married and so on, but they can't tell you about you.

Quantum mechanics has shifted a little bit the way we think about this fundamental philosophical point. Maybe the world isn't fully deterministic. Now, again you can ask the question, does learning that in the early 1900's throw away all of these wonderful ideas of classical physics? My viewpoint is that is not even remotely the case. Classical physics describes the world we live in as accurately today as it did in the 1600's when Isaac Newton was first making sense of it. Again, it's the underpinnings down at these levels that are far, far beyond our ordinary experiences, and it's useful to know about it. It opens up new branches of physics when you start asking these questions. Classical physics will not explain or predict a laser beam. A laser beam really arises from the quantum mechanics of electrons inside of an atom.

On the other hand, classical physics does a perfect job of explaining how that laser beam behaves. It behaves just as Isaac Newton thought. It goes through lenses. It focuses. We understand the color. We understand how to build a CD player with laser beams or the scanner at the grocery store with laser beams because ultimately the macroscopic phenomenon of laser light is just classical physics again. It is predictable and understandable. It is only when you ask that deep, deep question of where did it come from that you might need and want something new.

In this respect, modern physics has built on classical physics. It doesn't throw it away. It enriches it. It deepens it. It starts to take that framework, that cathedral, and it now builds something underneath it, some even deeper layer still. We may not be done. Classical physics continues to describe aspects of the practical world. Quantum mechanics continues to be studied and deepened. It's not necessarily clear that there is an end point. Classical physics has led us to believe that science is an investigative process. We're trying to make sense of the world that we live in. It is absolutely in my opinion the most fruitful, powerful, and productive discovery of ideas in the history of human civilization. It has had enormous impact on the world that we live in, the technological world that we live in, and all of the aspects of our lives are built upon and center on classical physics. We study it not because we're interested in history but because it really is telling us something very useful about the world that we live in and the things that go on around us. If you want to understand how a baseball behaves in a baseball stadium or how the electricity for your house is generated, or how your microwave oven works, these are ideas that can be understood from classical physics. It's a tool. If you leave this course and all of the sudden you realize that there are some important political issues involving energy and the environment, you should recognize that the tools of this course are sufficient for you to understand the scientific questions. Now, that's only one piece of the story. You have to think about the politics and the ethical issued involved and classical physics may or may not be helpful for you to move in those directions. It is enormously helpful for you to be able to separate science from wishful thinking as well inaccurate thinking if you're thinking about limitations in addition to environmental consequences, energy, conservation of energy, the definition of power and its relationship to energy. It's very, very fruitful.

My hope is that by the end of this course, the energy you've put into it has changed a little bit the way you think about the experiences that you have. Go out and think again when you're driving your car, when you're walking, when you're riding your bicycle, when you're looking at toys, when you're just thinking about the world around you. I hope that this inspires you to continue learning. There is an awful lot more that you can continue to study in the field of classical physics alone, and I thank you. It's wonderful when people make the effort to make sense of the world they live in.

Timeline

c. 500 B.C. .... Pythagoras founds his school on Samos. It holds as a basic belief that reality is fundamentally mathematical in nature.

384–322 B.C. .... The life of Aristotle, whose model for the motion of bodies was the standard for European science for almost 1500 years.

c. 150.............................. Ptolemy proposes the "epicycle" model of planetary motion in the *Almagest*, which stands, along with Aristotelian physics, through the Middle Ages.

1543 .... Nicolaus Copernicus publishes *On the Revolutions of Heavenly Spheres*, the culmination of his heliocentric theory. He proposes an alternative to the Ptolemaic model of the solar system, placing the Sun at the center, rather than the Earth. His work is very careful to steer clear of confrontation and controversy, avoiding censorship by the Church. Copernicus receives one of the first printed copies on his deathbed.

1570 .... Tycho Brahe constructs his observatory at Hven, which collects the data that Johannes Kepler uses to formulate his three laws of planetary motion. These data, predating the use of telescopes for astronomical observations, are of unprecedented breadth and precision.

1619 .... Kepler publishes *Harmonices Mundi*, which puts forth his three famous laws. These laws were deduced from reams of data collected over many years by Brahe, combined with years of intensive work and tremendous insight into geometry and mathematics.

1632 .... Galileo writes *Dialogue Concerning the Two Chief World Systems*, in which he attacks both Aristotelian physics and

Ptolemaic astronomy. His tone is confrontational—the Church-backed Aristotelian and Ptolemaic models are given voice in this "dialogue" by the pedantic dullard Simplicio. Galileo is called to trial on suspicion of heresy, put under house arrest, and forced to recant the views put forth in *Dialogue* and refrain from further publishing for the rest of his life.

1669 ....Isaac Newton is made a fellow of Trinity College. He immediately begins research on the subject of optics, including reflection and refraction and their practical application to telescopes.

1679–1687 ....Newton begins studying mechanics, primarily focused on gravity and orbital motion. This work culminates when he publishes the *Principia* in July of 1687, which lays out the groundwork of modern physics in his three laws. These would survive essentially unaltered for 200 years. Newton also lays out the law of gravitation, which successfully predicts and explains Kepler's three laws of planetary motion. Newton develops the mathematics of calculus (though Gottfried Leibniz developed calculus independently at the same time and began publishing his results sooner).

1696 ....Newton takes up a post as warden of the Royal Mint and is promoted to Master of the Mint upon his supervisor's death in 1699. He would do relatively little physics for the rest of his life.

1714 ....Leibniz develops a mathematics of motion based on energy, rather than momentum. His work is largely buried under nationalistic concerns (Descartes in France and Newton in England both focused on momentum), but later problems prove much easier to solve using the idea of conservation of energy; thus, both momentum and energy were eventually adopted as complementary approaches.

1733 .... Charles du Fay determines that electrical charge appears to come in two flavors, "vitreous" and "resinous," later renamed to *positive* and *negative*. He also finds that any substance could be charged by heating or rubbing it, except for metals and soft/liquid bodies. Furthermore, he discovered the basic rule of electrostatics: Like-charged bodies repel; oppositely charged bodies attract.

1742 .... Anders Celsius, a Swedish astronomer, proposes a new scale for temperature based on the freezing and boiling points of water.

1750 .... After proposing that du Fay's "vitreous" and "resinous" fluids are not, in fact, separate, but really separate manifestations of the same fluid (a step in developing what we now understand as electric charge), Benjamin Franklin proposes his famous kite experiment, to prove that lightning storms are caused by electrical forces.

1757 .... James Watt creates his steam engine, offering a considerable increase in efficiency over previous models. It is the industrial drive to seek ever more powerful and efficient heat engines that pushes much of the study of thermodynamics through the 19th century.

1779–1783 .... Antoine Lavoisier isolates and identifies the element oxygen. He then uses this to debunk the phlogiston theory of combustion. Lavoisier goes on to propose the law of conservation of mass.

1785 ....Charles Coulomb proposes an inverse-squared law for electric force and proves his theory by careful use of torsion balance (basically a spring scale). The unit for charge is named in his honor. He also determines some of the relationships of forces between magnetic poles but categorically refuses to accept the idea that any connection between electric and magnetic forces could exist.

1788 ....Joseph Lagrange develops Lagrangian mechanics as the culmination of work over 16 years to simplify formulas and ease calculations. He is arguably the greatest mathematician of the 18th century, who made his claim to fame based on his work on wave propagation and analytical mechanics, both of which are extremely useful to physics.

1798 ....Henry Cavendish determines the mass of the Earth and, by doing so, calculates Newton's gravitational constant, *g*. (He did this through the use of a delicate apparatus developed by John Michell, who died and left the instrument to Cavendish.)

1800 ....Alessandro Volta invents the prototype battery (the *voltaic pile*). Prior to this development, all charge used for experimentation came from *Leyden jars*—capacitors that could provide a burst of electrical current. Volta's voltaic pile allowed the study of steady electrical currents.

1800 ....John Dalton becomes secretary of the Manchester Literary and Philosophical Society, through which he eventually publishes his atomic theory: All matter is made out of small, indivisible atoms.

1801 ....Thomas Young conducts his double-slit experiment. A beam of light is passed through two narrow slits, creating a diffraction pattern on a screen behind the slits. This is strong evidence for the wave nature of light.

1811 ....Amadeus Avogadro proposes Avogadro's law: that containers (at the same temperature and pressure) of different gases contain the same number of molecules, regardless of the chemical or physical properties of the gases.

1819 ....Hans Oersted discovers (possibly by accident, while preparing for a public lecture) that a wire carrying a current will divert a compass needle. This discovery provides the starting point for discovering the connections between electric and magnetic forces that would ultimately culminate in James Maxwell's work.

1820–1826 ....Andre Ampere develops the mathematical representation describing the Oersted discovery. This representation explains magnetism as resulting from the motion of many small charges.

1821–1831 ....Michael Faraday discovers the dynamo principle and demonstrates electromagnetic induction. Toward the end of this time, he begins to finalize the idea of a field—a concept whose mathematical expression would culminate in Maxwell's equations.

1824 ....Sadi Carnot formulates the idea of the *Carnot engine*, a steam engine of theoretically ideal efficiency. He uses this as a thought experiment to prove that temperature is the most important variable in an engine, not the material or specific construction details.

c. 1845............................Henry Joule discovers and refines the principle that comes to be known as *Joule's law*—the conversion of mechanical work to thermal energy. In this year, Faraday also begins corresponding with William Thomson, who begins the initial efforts on mathematically expressing the ideas of Faraday's fields before passing the project along to Maxwell.

1847 ....Hermann von Helmholtz proposes the law of conservation of energy, the first law of thermodynamics, as a development of medical studies of muscles. He expands this, connecting heat, motion, magnetism, and electricity as various forms of energy.

1861–1868 ....James Clerk Maxwell unifies electricity and magnetism in a series of papers and proposes the electromagnetic nature of light.

c. 1881...........................Heinrich Hertz experimentally shows the existence of electromagnetic waves, providing the basis for radio technology, as well as proving Maxwell's equations.

1882 ....J. Willard Gibbs begins publishing his work on statistical mechanics.

1884 ....Ludwig Boltzmann develops a theory of blackbody radiation, deriving from statistical arguments the empirical relationship that had been discovered by Josef Stefan. He independently develops much of the same theories that Gibbs did.

1900 ....Max Planck publishes his theory of blackbody radiation. He builds strongly off of Bolztmann's statistical physics but introduces the requirement that the energy of photons must be contained in discrete bundles.

1905 .... Albert Einstein's "Miracle Year." He publishes three papers, any one of which would be enough to cement his place in science. All three in a single year make him a name for the ages and demarcate the end of the era of classical physics and the start of modern physics.

Glossary

AC (alternating current): Electrons in a circuit oscillate back and forth instead of flowing (compared with DC, or direct current).

acceleration: The rate of change in velocity; defined as change in velocity divided by time passed. Because velocity is a vector, so is acceleration (and we *can* have acceleration at constant speed if the *direction* of velocity changes!).

acoustics: The branch of physics that studies sound.

action at a distance: A property of many early theories (including Newton's theory of gravity or Coulomb's electric force law) stating that distant objects affect each other.

amp or ampere: The metric unit of current; it indicates the number of coulombs flowing past a given point each second.

angular momentum: A quantitative measure of how rapidly objects are turning coupled with how massive they are and how that mass is distributed. Angular momentum is a quality of any spinning object and is conserved (provided that the object is not subject to an external "twisting force," or torque).

angular speed: The rate at which something is spinning; can be measured in revolutions per second or radians per second. This is related to angular momentum but without regard to the mass or distribution of mass.

arrow of time: An abstract idea that dictates which processes are reversible, such as a billiard-ball collision, and which are irreversible, such as cracking an egg.

atom: Originally, the fundamental (indivisible) building block of all matter. Now, the smallest building block of chemistry, an individual particle of any element. Physically, a heavy nucleus with electrons orbiting.

atomic hypothesis: All physical matter is a composite of atoms.

battery: A mechanical device that turns chemical energy into electrical energy; if connected to a circuit, it will drive a current at a given voltage.

Brownian motion: The erratic motion of small but visible objects (e.g., dust) resulting from collisions with smaller (microscopic) atoms and molecules. Einstein's quantitative description of Brownian motion was the final piece of evidence in convincing physicists of the physical reality of atoms.

caloric theory: An early and now discredited theory of thermodynamics stating that heat is a physical fluid, rather than a transfer of energy.

center of mass: An average position of matter in an object; the effective point where gravity (or external force) acts.

charge: A property of all matter that determines electrical forces. Charge can be positive, negative, or neutral. Electric field lines start on positive charge and end on negative charge.

chemistry: The study of combinations of atoms and the resulting compositions and combinations of matter.

circuit: An electrically conducting path that can carry current in a loop.

classical physics: Basically physics before 1900; characterized as deterministic and realistic. Classical physics includes kinematics, mechanics, optics, thermodynamics, electricity and magnetism, and more (acoustics, fluid dynamics,...).

conduct: To allow charge to flow in a material. (A *resistor* still conducts, just with more resistance. The opposite of a conductor would be an insulator.)

conservation law: The situation that exists when some quantity remains unchanged during an interaction. For example, charge conservation states that the sum of all electric charges never changes in any particle reaction. Energy conservation states that although energy may transfer from particle to particle or form to form, the total (numerical) sum remains unchanged.

constructive interference: A defining property of waves in which two like waves add together, "building up." Note that if they are traveling waves, they then continue without affecting each other further on.

cosmology: The field of physics that studies the history, structure, and evolution of the universe.

coulomb: The unit of electric charge. One coulomb of charge is a *lot* of charge! (An electron carries 1.6×10^{-19} coulombs.)

Coulomb's law of static electricity: An equation that predicts the force between any two stationary charges at a given distance. Force is proportional to $Q_1Q_2/\text{distance}^2$.

current: The flow of electric charge (measured in amps).

DC (direct current): Electrons in a circuit flow in only one direction. (Compare with AC, or alternating current.) DC would result from a circuit with a battery; AC would result in household circuits.

degrees of freedom: A measure of the possibilities for the shape and location of an object. More degrees of freedom offer more possibilities to move and change. For example, a particle stuck on a rod can move only back and forth, restricting its degrees of freedom. In statistical mechanics , degrees of freedom of the pieces are very important in computing the entropy of a system.

destructive interference: A defining property of waves in which two like waves, out of phase, can cancel each other (add up positive and negative to give zero displacement) at one point.

determinism: A philosophical belief that if all physics and the exact state of the universe could be known at any given time, then the future could be perfectly predicted.

diffraction: The bending of light around corners.

dynamics: The branch of physics dealing with the "why" of motion. (Newton's laws are about dynamics of particles. Thermodynamics explains the flow of heat.) Compare with simple descriptions or, for example, kinematics.

E & M: The field of physics that studies the fundamental forces of electricity and magnetism, their sources, and their connections.

efficiency: Useful energy output divided by total energy input for a machine. (As we stated it in the course: "what you get" divided by "what you paid for.")

electric field (or e-field): A property of a location in space that indicates the force an electric charge would feel if placed at that location.

electricity: The forces and fields that result from the interactions of charged particles.

electromagnetic field: Maxwell's field that simultaneously describes electric fields and magnetic fields and their interactions. It is a unified, more universal way of thinking about electric and magnetic fields together. Light is an electromagnetic field—both electricity and magnetism are required to make sense of the phenomenon; they are intimately connected to each another.

electromagnetic wave (or EM radiation): The unique self-propagating wave of electric and magnetic fields. The only known wave that does *not* require any physical medium to propagate. At the right frequency range, this is commonly called *light*. At other frequencies, it includes (in increasing energy, which is also increasing frequency, but decreasing wavelength): radio waves, microwaves, infrared radiation, light, ultraviolet rays (UV rays), x-rays, and gamma rays.

ellipse: A particular kind of stretched circle; the path of planets in orbit. Mathematically, one of the *conic sections*.

energy: A measure of the amount of work (as defined by physics) that an object can do, at least in principle.

entropy: A quantitative measure of the disorder of a thermodynamic system.

equilibrium: A state of balance; a system or interaction of systems in which nothing macroscopic is changing.

experiment: A controlled test or procedure, often one that compares the predictions of a theory with the behavior of the universe.

experimentalist: A scientist who primarily devotes his or her time to creating and running experiments to better understand nature (as compared to a theorist). This is a more modern specialization/distinction; in the classical physics era, many scientists played the roles of both experimentalist and theorist.

field: The physical manifestation of a force of nature, present throughout space. An alternative way of thinking about forces, rather than "action at a distance."

field lines: A visualization tool to picture how electromagnetic fields appear in the presence of sources (charges or currents). A *field line* shows which way test charges would *start* to move if released at that point (tangent to, or "along," the field lines). Where the field lines bunch together, the forces are strongest.

force: A push or pull on an object; this is what causes acceleration. Force has a strength and a direction. Force is the quantitative measure of interactions of objects.

force field: The idea that a source of force produces something (a field) in all of space, whether or not there is an object there to feel it. For example, the Earth produces a *gravitational force field* around itself.

freefall: The idealized motion of an object falling due to gravity, without air resistance or any *other* forces. (A parachutist is in true freefall only briefly: Air resistance builds up quickly, ultimately becoming just as important as gravity, at which point the parachutist falls at constant speed.)

frequency: A measurement of how many times an object oscillates in a second; it is measured in hertz, or Hz. Thus, 60 Hz means "60 cycles each second."

frictionless: An approximation commonly made to simplify physics questions, in which we neglect the (often small but rarely truly zero) effects of air resistance or surface resistance to motion.

fundamental forces: The four basic forces that cause all motion and bind together all matter; they are the gravitational force, electromagnetic force, strong force, and weak force. (The latter two are not part of the classical physics story.)

gamma rays: High-energy EM radiation; see **electromagnetic wave**.

geocentric ("Earth centered"): The belief that the Sun and the entire universe rotated around the Earth.

gravity: One of the four fundamental forces. A purely attractive force generated between any two masses; it depends on the distance between the masses. Newton's formula of universal gravity is expressed as follows: Force is proportional to $M_1M_2/\text{distance}^2$.

ground: Electrical term referring to any object big enough to give or receive charge without itself becoming electrically charged; usually the Earth.

heat: A verb describing the transfer of thermal energy into or out of a system.

heliocentric ("Sun-centered"): The belief that the Earth, along with the rest of the solar system, revolves around the Sun.

hertz: The unit of frequency; it indicates number of cycles per second.

inertia: The tendency of an object in motion to remain in the same motion; also, a quantitative measure of the resistance of any object to change in its velocity (for a given force). Mass is the direct measure of inertia.

infrared radiation: See **electromagnetic wave**.

insulate: To prevent the flow of electrons through or on a material. Many materials (wood, plastic, and so on) are good insulators. Insulators can "break down"; for example, a high-voltage source can cause a spark, which means a conducting path has been created in what was previously an insulator.

interaction: A synonym for *force*; a way in which particles transform or perturb one another.

internal forces: Forces between objects *inside* a system. (Distinguished from *external forces*, which arise from something *outside* a system.)

invariant: Any property that remains unchanged over time.

kinematics: The study of physics in which motion is described. Kinematics of particles involves the relationships among position, velocity, and acceleration over time.

kinetic energy: The energy an object has based purely on its speed. The classical formula for kinetic energy is $(1/2)mv^2$.

law: A fundamental theory on which many other theories are based; a pattern or relationship that is extremely well established experimentally. However, even a “law of nature” may not be true in certain limits. For example, Newton’s law of gravity must be modified in extreme situations, such as near a black hole. Newton’s second law must be modified at velocities near the speed of light or for subatomic particles.

light: Visible electromagnetic radiation. See **electromagnetic wave**.

macroscopic: Anything of “human scale”; bigger than what can be seen with a low-powered microscope; very roughly larger than micrometers.

magnetic field (B field): A property of a location in space that indicates the force a moving electric charge (or a permanent magnet) would feel at that location.

magnetic poles: The magnetic analogy to charge; the points on a magnet where the field lines enter or exit the magnet. (There are north and south magnetic poles.)

magnetism: Another fundamental of force of nature; also, the study of magnets, currents, and the interactions among them.

mass: A measure of how much “stuff” an object has; the measure of inertia of a body; the quantity that determines the gravitational force on an object.

matter: A generic term for everything made of atoms (or the material components of the world). Matter has mass.

medium: The background material that supports a wave. For example, water is the *medium* for ocean waves; air is the *medium* for sound waves.

microscopic: Referring to a scale smaller than the human scale; invisible to the naked eye. Roughly, the molecular scale; certainly smaller than a fraction of a micrometer. Often synonymous with *atomic scale*.

model: A simplified way of thinking about or picturing the workings of a complex system. For example, a solid object can be modeled as a number of microscopic hard spheres, connected by a grid of simple springs. A model need not always be literally correct, but it allows scientists to make predictions and understand systems.

modern physics: Physics from 1900 until today, which primarily deals with the very small (quantum physics) and the very fast or very massive (relativity). Modern physics has branched out to include studies of particle physics, plasmas, cosmology, lasers, and much more.

molecule: A chemical building block that is not fundamental but is built up out of a bound state of two or more atoms; for example, an H_2O (water) molecule.

momentum: A measurable property of objects (or systems). Related to the tendency of an object to continue in its motion; the "oomph" an object would have if it hit you. (*Force* tells you the rate at which momentum changes.) Momentum is *defined* for a particle to be mass × velocity.

monopole: A beginning or end of field lines (e.g., a positive charge is always at the beginning of electric field lines). Important because no one has ever found a magnetic monopole; therefore, magnetic field lines can never begin or end—they must form loops

motion: The description of the change in position of an object over time, measured by velocity and acceleration. "At rest" is a state of motion in which the object has zero velocity and zero acceleration; its position is not changing.

neutral objects: Objects with equal positive and negative electric charge, thereby appearing to have no charge at all.

Newton's second law (N-II): The heart of dynamics; the law of nature that says that force causes any mass to accelerate according to the formula $F = ma$, or as Newton would write it, force = (change in momentum)/(time taken). The formula is a *vector equation*, meaning that the *direction* of the force tells you the *direction* of the acceleration.

nucleus: A bound collection of protons and neutrons; the center of atoms.

Ockham's razor: A scientific principle basically stating that if all else is equal, then the simplest theory is more likely to be right.

optics: The study of light and how light travels through and between materials. (Geometric optics thinks of light as *rays*; physical optics tends to think of light as *waves*—both can be important!)

parabola: A mathematical shape that describes the path of a thrown object; commonly seen in archways. Mathematically, one of the *conic sections*.

particle: A discrete bit of matter; an idealization or simplification used to think about objects whose internal structure is irrelevant. (A *fundamental particle*, such as an electron, has mass but no relevant volume.)

particle physics: The study of the fundamental constituents of nature and their interactions with one another.

periodic table: Dmitri Mendeleev's organization of atoms into a simple table, in increasing order of weight, which shows the underlying structure of atoms.

physics: The science of the physical, measurable world. The study of matter and energy, space and time, and the structure and interactions of things in the world.

position: The place an object is located at a given time.

potential energy: Energy that is stored in an object to be released later (e.g., the chemical potential energy of gasoline); commonly refers to the gravitational potential energy an object has simply by virtue of being a height off the ground. Potential energy arises from interactions; for example, gravitational potential energy is a manifestation of the force of gravity and the work it can and will do if objects are allowed to fall toward one another.

pressure: Force per unit area.

quantum mechanics: The physical theory that tells how microscopic particles behave (as contrasted with classical or Newtonian mechanics, which is based entirely on force and Newton's laws).

radio waves: See **electromagnetic wave**.

reductionism: The philosophical principle that complex systems can be understood once we know what they are made of and how the constituents interact.

reference frame: The perspective from which a person makes measurements. An *inertial reference frame* is one in which Newton's first law holds.

relativity: A deep principle of physics that says that the laws of physics are the same in every inertial frame. Galileo postulated this, but the generic term is now usually reserved for Einstein's theories, which describe the motion of particles moving at high speeds. Special and general relativity modify our conventional views of space, time, and gravity (but do *not* say that "everything is relative").

resistor: An element in a circuit that reduces the current that can pass for a given voltage. The filament in a light bulb is a good example of a resistor.

rotational motion: See **angular speed**

special relativity: See **relativity.** Einstein's 1905 theory describing the motion of particles moving at high speeds. (*Special* means that the theory is limited to observers moving with steady velocity and ignores gravity.)

speed: A measure of the rate at which an object's position is changing per second; how quickly an object is moving. (Distinguished from *velocity*, speed is just a number; it has no direction associated with it.)

stability: A qualitative measure of how difficult it is to change the state or orientation of an object (e.g., a pencil standing on end is very unstable, while a pencil laying on its side is more stable).

static electricity: The study of electric effects arising from charges that are not moving.

static equilibrium: A state in which an object is subject to zero net force (and zero torque or "twisting") *and* does not feel a large or increasing force if it is moved. Thus, a balanced pencil is in equilibrium, but the equilibrium is not *static* because the slightest perturbation will remove the pencil from equilibrium.

superpose: Adding two forces or fields according to the rules of vector addition. Two opposing forces *superpose* to yield a total force of zero. Two parallel forces *superpose* to yield a doubly strong force.

system: Any group of objects under consideration.

temperature: A measure of the average thermal energy an object has. From a more practical perspective, temperature is what thermometers measure!

test particle: A theoretical particle of infinitely small mass and charge used to map out force fields (without itself changing the field).

theorist: A scientist who primarily devotes his or her time to studying the mathematics and concepts in current physics to extend the limits of those theories (as contrasted with an experimentalist).

theory: A well-tested, organized way of understanding a broad variety of physical circumstances. It is not an idle "guess" or speculation (as the term is sometimes used in standard English); physicists do not use this word to mean "I have a theory about what's going on here." Examples include Newton's theory of gravity or Einstein's theory of special relativity. These are mathematical and physical formulations that organize and consolidate vast amounts of data. Theory combines facts, laws, and tested hypotheses.

thermal energy: The total energy an object has stored internally, in the form of kinetic energy of vibrations of its atoms.

thermodynamics: The study of temperature, heat, and thermal energy.

UV or ultraviolet radiation: See **electromagnetic wave**.

unification: The goal of physicists to find a deep connection between forces. Electricity and magnetism were unified by Maxwell in the 1860s; they are both manifestations of one underlying *electromagnetic field*.

vector: A mathematical arrow that contains information about the direction and size of a quantity. Examples of vectors include velocity (which is not just speed but also direction) or force (which is not just how hard a push is but also which direction it is in). Contrast this term with a *scalar*, which is just a number, for example, temperature or mass.

velocity: The rate at which an object's position is changing per second; how fast and in what direction an object is moving. (*Speed* is the magnitude of the velocity.)

volt: The unit of electric potential in a circuit; a quantitative measure of the electrical potential energy per unit charge. Voltage differences tell, very crudely, the amount of "pressure" felt by electric charges in a circuit.

wave: A self-propagating disturbance (usually of some medium, except for EM waves) that can carry energy but is not itself a particle.

wavelength: In a wave, the distance between repeating parts of the wave.

work: The physics term for the result of a force pushing (or pulling) on an object over some distance. Work is form of energy transfer. The definition is: work = force × distance traveled, where one counts the component of force only in the direction of motion.

X-rays: High-energy electromagnetic radiation. Often used as a synonym for *gamma rays*, although X-rays connote slightly lower-energy radiation. See **electromagnetic wave**.

Biographical Notes

A biographical list of "key contributors" to the development of classical physics is almost impossible to compile because the number of contributors is so large! Although famous physicists often get sole credit for their accomplishments, the great discoveries are inevitably part of a web of scientific progress. Truly significant contributions come from both brilliant and more mediocre scientists, not to mention support from graduate and sometimes undergraduate students, technicians, lab assistants, and so on. Most of the discoveries in physics have a complex lineage; historians could (and do) quibble about the attributions and origins of almost every idea in the field. Some ideas get "rediscovered" or further developed, then attributed to the scientist who somehow was better able to spread the word. What follows is an extraordinarily abbreviated list of some of the *most* famous names in the field. The large number left out is painful to this "biographer" (who is also admittedly not even remotely a historian of physics).

It is also worth noting that the physics described in this course is directly descended from the Western European tradition, and it may seem culturally insensitive not to mention discoveries made by Asian, Native American, African, or Arabic science. This arises from a combination of the shameful ignorance of the author of this text and the nature of contemporary Western science education, in which there is a direct pedigree and narrative that flows through the Western tradition. By no means should we discount the discoveries that were made on other continents, but the historical story that sits aside conventional (Western) introductory courses in physics is predominantly European in nature.

Greek Philosopher-Scientists (600 B.C.E.–A.D. 150)

Western philosophy, including *natural philosophy* (the philosophical precursor to science) traces its roots back to the Greeks. Because this course deals with the Western tradition of physics, we need to mention the beginnings, even if only in passing, because it was the growth away from the ideas of the Greeks that has largely defined physics as we now know it.

Aristotle (384–322 B.C.E.). Aristotle's name is perhaps most important to modern science, not for the work he did, but as a symbol. Up until the 18th and early 19th centuries, the studies of philosophy and science were closely linked. Aristotle was the author of the philosophical and scientific system that was to dominate Western thought into the 17th century. Although he has a large body of work devoted to mathematics, philosophy, and zoology, the most pertinent of Aristotle's works for our course is his view on the motion of physical bodies.

Aristotle maintained that the speed of an object is determined by the magnitude of the force pushing it: The greater the force, the faster the body will move. Although initially quite plausible (and correct in situations where viscous drag dominates; indeed, this is one of the most common mistakes made by students in introductory physics courses!), this model breaks down when applied to astronomical bodies. It was the interpretation of the stars (first by Ptolemy, followed by Copernicus, Kepler, Galileo, and finally, Newton) that proved to form a more robust basis for the study of physical motion.

Aristarchus (310–230 B.C.E.). Aristarchus was a mathematician and astronomer who came from the same island as Pythagoras. Most of what survives of his work comes in the form of quotes from other writers of his era and a few indirect references. Still, he is often credited with being one of the initial proponents of a heliocentric model—one in which the Earth and other planets rotate about the Sun. The only surviving work of Aristarchus is *On the Sizes and Distances of the Sun and Moon*, which attempts to estimate the distances between the Earth, the Sun, and the Moon by cunning use of geometry.

Ptolemy (c. 85–165). Ptolemy was the proponent of the geocentric (Earth-centered) model of the solar system that was to prevail in Europe for 1400 years. Unlike those of most other classical authors, many of Ptolemy's works survived the trials of time and are available to us now in their original forms. Based on philosophical and scientific ideas, Ptolemy's model proposed that the Sun and all the planets rotated about the Earth, each having a complex system of "gears" (epicycles) that rotated at the same time in order to account for the motion that the planets exhibit in the nighttime sky. (The word *planet* is derived from the Greek word for "wanderer.") The Ptolemaic model of the solar system is impressively accurate, satisfying observational astronomy's needs until the "modern" high-precision observations of the late 16th and early 17th centuries.

The Birth of Classical Physics (1500–1800)

Nicolaus Copernicus (1473–1543). Copernicus was born in Poland, the son of a copper merchant. When Nicolaus was 10, his father died, and he and his siblings were placed in the custody of their uncle, a Church canon. Nicolaus was educated at the cathedral school, then enrolled in the University of Krakow, where he received an education in Latin, philosophy, mathematics, and (most importantly) astronomy. He became a canon, like his uncle, and continued his education throughout his life, adding Greek, medicine, and canon law to his list of accomplishments.

Copernicus is given credit for one of the first proposals of a Sun-centered universe since the classical age. He first proposed his ideas in a small handwritten book in 1514 that he circulated anonymously amongst his friends. Meanwhile, his other duties included advising the pope on calendar reform, organizing the defense of his hometown, and carrying out currency reform. It wasn't until the end of his life, in 1543, that he published his final theory of the heliocentric universe. His magnum opus was deliberately circumspect, presenting itself as a theory rather than the absolute truth and, thus, avoiding controversy and Church censorship. Although the details were still wrong (Copernicus assumed perfectly circular orbits for the planets), Galileo, Brahe, and Kepler all read this work and were profoundly influenced by it.

Tycho Brahe (1546–1601). Tycho Brahe was born into the Danish nobility; both his father and mother came from important families that were influential with the Danish king. At the age of 13, he began attending the University of Copenhagen, ostensibly to study law, but he soon discovered that his real passion was astronomy (helped along by an eclipse in 1560). As part of his studies, Brahe traveled on the Continent. In 1567, he was involved in an altercation with another Danish student while he was in Germany that resulted in part of his nose being cut off in a duel.

By the 1570s, Brahe had earned a reputation as Denmark's preeminent scientist. When he announced his intention to leave Denmark, the king offered him an island on which to build a royal observatory as enticement to stay. This observatory at Hven collected data on the position of stars in the sky to an unmatched degree of precision (without the use of telescopes, which were not yet in use for astronomical observations!). Brahe would pass on these data to Kepler, leading to Kepler's three laws of celestial motion. In 1588, the king who had appointed Brahe died, and Brahe was not in favor with the new monarch. He left his position in 1597 for Germany, where he was appointed imperial mathematician in 1599.

Brahe died at the age of 65. The (possibly apocryphal) story is that he died from a ruptured bladder, caused by his refusal to leave the table at a feast before his host did.

Johannes Kepler (1571–1630). Kepler was the son of a mercenary soldier and an innkeeper's daughter in what is now modern Germany. His father died in a war in the Netherlands when Johannes was 5 years old. The boy was schooled by monks and, throughout his life, was profoundly religious, seeing his work as part of the Christian duty to understand the works of God.

The contribution for which Kepler is most remembered is his development of the three laws of planetary motion. As imperial mathematician, Kepler inherited all the data collected by Tycho Brahe over the previous 40-year period. Kepler, through grueling hours of analysis (over the course of years), deduced the motion of the planets and discovered that it fit most accurately with the Sun-centered model of the solar system put forward by Copernicus, albeit with the critical modification that the planetary orbits are not ideal circles but elliptical paths. Furthermore, Kepler's analysis of the planets' motion in time allowed him to state three geometrical laws that appeared to describe the motion of *all* planets. It was the crowning achievement of Newton's law of gravitation that it was able to re-create and derive Kepler's observational laws from first principles.

Galileo Galilei (1564–1642). Galileo was the son of a musician living near Pisa, Italy. At a young age, he sought to join a monastic order but was forced to return home by his father, who had already decided that his son was to become a medical doctor. Over the course of his medical studies, Galileo was exposed to mathematics and natural philosophy, which he took to immediately. He slacked off in his other classes, devoting all his energy to those two subjects. By the age of 21, he had earned an appointment as a teacher of mathematics in Siena. By 1592, he was offered an appointment as a professor of mathematics at the University of Padua.

Although much of Galileo's work was in the area of engineering (he more than doubled the magnification of telescopes of the time) and astronomy (it is for him that the Galilean moons of Jupiter are so named), the work for which he is most famous is the *Dialogue of Two Chief World Systems*, which he published in 1630. This was framed as a debate between the Copernican model of the heavens, which placed the Sun at the center and the other planets orbiting around it, and the Ptolemaic system, which placed the Earth at the center of the heavenly system. The confrontational manner in which he wrote this work earned Galileo censure by the Church.

Galileo is also perhaps the most famous figure associated with overturning the Aristotelian and Ptolemaic models of physics, in large measure because of his conflict with the Church. His work was extremely insightful, however, and inspired many future scientists, up to and including Einstein.

Isaac Newton (1643–1727). Newton is the father of physics, indeed, of all of modern science in many ways. I can think of few individuals of the last 1000 years with more direct and profound influence on the human condition. Isaac Newton's masterwork, the *Principia*, articulated not only a number of physical laws but also the scientific method itself. Newton's laws describe and (to some extent) explain motion and gravity. When faced with the need to solve the equations he developed, Newton *invented* the calculus required to solve them. His central laws are *universal*, applicable to any system in any circumstance. Even today, their accuracy and power is extraordinary. Although Newton's laws must be extended under extreme conditions (for example, for objects traveling near the speed of light), they still form the basis for much of modern technology. Newton was involved with both theory and experimentation, and his research touched on and formed the roots of many branches of modern physics, including optics, thermodynamics (heat), fluids, and more. Students in freshman physics learn about Newton's work in their first semester (then repeatedly, with further depth, as they progress). The metric unit of force, the *newton*, is named in his honor. Newton was not a pleasant or easy man. He had a big ego, never married, and had many disputes over intellectual priorities during his life. However, in an uncharacteristic but famous quote, he said, "If I have seen further, it is by standing on the shoulders of giants."

Henry Cavendish (1731–1810). Of all the 18th-century physicists, Cavendish was one of the most eccentric. He was pathologically shy—to the point of having a second staircase installed in his house so that he could avoid seeing his housekeeper. When invited to scientific salons (a fashionable dinner party featuring important figures in a field of literature or science), the only social event he would attend, people seeking his opinion were advised to enter the room, not look at him, and speak their questions to the opposite wall.

Because of his shyness, Cavendish rarely published, and later review of his work revealed that he had discovered many laws relating to properties of gases, chemistry, and electricity before others who were credited with their discovery. The published work for which Cavendish is best remembered (*Philosophical Transactions of the Royal Society of London*, 1798) was the extremely accurate and precise determination of the density of the Earth using an experiment designed by John Michell (who died before he could complete it). The experiment used a delicate torsional balance, as well as mirrors and optics, to measure the force between two large weights in the lab. The current results for the mass of the Earth deviate by only 1% from the results that Cavendish obtained more than 200 years ago. The result is now framed as the first measurement of the universal constant of nature (known as G), which appears in Newton's formula for the gravitational force between objects.

James Watt (1736–1819). Watt was born the son of a Scottish shipwright and was largely home-schooled by his mother. At the age of 17, he traveled to London to become an instrument maker and became the head of a small workshop at the University of Glasgow upon his return to Scotland. While at Glasgow, Watt became interested in steam engines (then in their earliest stages). After studying one that was in the possession of the university, he came up with a scheme to dramatically improve their efficiency, but the machining issues in creating a prototype, as well as finding funding for the effort, proved difficult, occupying Watt for nine years before he formed a partnership with Matthew Boulton.

Both Boulton and Watt made a fortune off the steam engines produced by their partnership. Watt is also given credit for inventing the steam locomotive in 1784. Although he was not a scientist by trade, it was Watt's inventions that drove an entire branch of scientific inquiry for the next 100. The improved efficiency of his new steam engine also opened the way for an industrial revolution powered by steam and coal, rather than by rivers and waterwheels.

Nicolas Sadi Carnot (1796–1832). Carnot grew up through the tumult of the French Revolution and the Napoleonic wars. He was home-schooled by his father in mathematics and science, as well as language and music. After studying at the École Polytechnique under such notables as Ampère and Poisson, Carnot enrolled in a two-year course in military engineering.

After leaving active duty in the military to attend more courses in Paris, Carnot began (in 1821) studying the mathematical theory of heat, which led to modern thermodynamics. He was driven by the problem of designing more efficient steam engines. His name is now attached to the *Carnot engine*, an idealized device that mathematically proved the theory that ideal efficiency of a heat engine depended on the difference in temperature between the engine itself and the surrounding environment, not on the nature of the substance used in the engine.

Carnot died in 1832 at the age of 36, only a day after contracting cholera in an epidemic that swept Paris.

James Prescott Joule (1818–1889). The son of a wealthy brewer, Joule had an early scientific education (home-schooled for 16 years, then tutored by John Dalton of chemistry fame). Joule was active in running the family brewery until its sale in 1854, initially treating science as a hobby. When he began looking into replacing the brewery's steam engine with a new electrical engine, science began to occupy more of his life.

Initially ignored by the Royal Society of London as a provincial dabbler, between 1840 and 1850, Joule discovered the law that is named for him—showing the connections between mechanical energy and heat—and continually refined his experiments, improving the accuracy of the results (because the nature of his discovery demanded extremely precise measurements!).

Although there was a dispute with a German scientist, Julius Robert von Mayer, as to who was the first to determine the relationships among work, energy, and heat, the unit that modern scientists use for energy is named after Joule.

The Early Chemists (1750–1860)

Antoine Lavoisier (1743–1794). Lavoisier was a French nobleman whose contributions feature prominently in chemistry, biology, finance, and economics. He identified the element oxygen in 1779 and showed that respiration by living beings was essentially the very slow combustion of organic material inside the body. In 1783, he dethroned the phlogiston theory of combustion, the previous model by which chemical reactions were understood.

Lavoisier introduced the law of conservation of mass; that is, that in chemical reactions, matter is neither created nor destroyed but simply changes form. His *Elementary Treatise of Chemistry* is considered to be the first textbook of modern chemistry.

His accomplishments in science aside, Lavoisier was a prominent member of the French nobility and served as a tax collector. The French Revolution did not treat such people well, and in 1794, he was framed for treason and guillotined (he was exonerated by the French government a year and a half after his death).

John Dalton (1766–1844). John Dalton was born in Cumberfield, England, and educated by his father, a teacher at the Quaker school in the same town. At the age of 12, he assumed that post upon his father's retirement. Although his initial foray into teaching was a disaster, he kept at it and passed the majority of his life earning a living as a teacher, either at a public post or as a private tutor.

In 1800, Dalton became the secretary of the Manchester Literary and Philosophical Society, through which he proposed the major work that he is remembered for. Inspired by Lavoisier's work, Dalton proposed his atomic theory, that is, that all matter is made of tiny, indivisible atoms. Everything about a certain atom can be known by knowing what element that atom is—all atoms of one element are identical but are fundamentally different from the atoms of each other element. Dalton's atoms can be neither created nor destroyed and are only moved about in chemical processes.

In 1837, Dalton suffered an attack of paralysis and again in 1838. In early 1844, he suffered a stroke, and his last meteorological observation is recorded the day before he was found dead by an attendant in July 1844.

Amedeo Avogadro (1776–1856). Born to a noble Italian family, Avogadro graduated with a law degree and began practice at the age of 20. Soon thereafter, he became interested in physics and mathematics and, in 1809, began teaching both subjects at the high school level. While he was teaching, he first proposed what is now known as Avogadro's law—that gases of equal temperature, pressure, and volume (no matter what the gas is) contain the same number of molecules. *Avogadro's number* is named in his honor and refers to the number of atoms contained in 1 mole of substance (roughly 602 septillion—602,200,000,000,000,000,000,000—a very large number!)

Relatively little is known about Avogadro's personal life. He was given a post at the University of Turin in 1820 as a professor of physics. Though he was restricted from teaching for a time because of his political support of Sardinian revolutionary movements, he ultimately taught there until 1853, with only a brief hiatus.

The Exploration of Electricity and Magnetism (1700–1900)

Benjamin Franklin (1706–1790). Franklin is probably one of the most well-known classical scientists, after Newton and Galileo. His fame, however, comes mostly from his nonscientific pursuits (publisher, ambassador, revolutionary). In 1748, 15 years after publishing the first *Poor Richard's Almanac*, Franklin retired from the printing business to pursue other opportunities, scientific experiments being chief among them.

Franklin's most famous work was in the field of electricity. Scientists before Franklin had identified two different types of "electrical fluid," the "vitreous" and "resinous." Franklin was one of the first to propose that there weren't two separate fluids but that both were simply different manifestations of the same fluid—a concept we know today as electric charge. His famous kite experiment (whether Franklin ever actually performed it is uncertain) was designed to prove that lightning was electrical in nature by flying a kite in a storm and showing that it collected an electric charge. The practical application of this theory led Franklin to invent the grounding wire and the lightning rod. Franklin's writings were well received in Europe, and his fame as a scientist overseas played no small role in his becoming an ambassador to Europe for the colonies before and during the American Revolution.

Charles-Augustin de Coulomb (1736–1806). Charles Coulomb's father was a successful lawyer and administrator, while his mother came from a quite wealthy family. As a child, he was given the finest classical education, studying language, literature, and philosophy, in addition to more modern subjects, including mathematics, botany, chemistry, and astronomy.

Coulomb studied to be an engineer for the French army, becoming an expert in structural design, fortifications, and soil mechanics. While working as a military engineer, he wrote seven papers (between 1785 and 1791), in which he developed the theory of attraction between charges, describing both how the force decreased with distance and how positive and negative charges interacted. Though these were his most important works as far as posterity is concerned, Coulomb participated in more than 300 committees for the French Academy of Science and wrote 25 memoirs, in addition to collaborating with many other important French scientists of his era.

Coulomb survived the tumult of the French Revolution (including the dissolution of the Academy of Science and its re-creation as the French Institute) and spent the twilight years of his life as inspector general of public instruction, setting up schools across France.

Alessandro Volta (1745–1827). Volta grew up and was educated in Italy. He was a professor of physics in Lombardy for most of his life, primarily interested in the study of electricity. In honor of his contributions to science, he was given the title of count by Napoleon in 1810. It is for him that we name the unit of electrical potential energy (per unit charge), the *volt*.

Volta's most famous contribution to science was the development of the *voltaic pile*, a very early battery. Previously, electricity had been studied by building up charge on metal spheres (in the same way that scuffing your feet across the carpet on a dry day builds up static charge on you), then studying how it interacted with other charges. Volta's chemical battery allowed for a steady source of charge to be produced, making it possible to study charges in motion and at rest. Coulomb turned electric forces into testable laws in the same way that Newton and Cavendish had done for gravity. The voltaic pile paved the way for Oersted and Ampère to do the same for electrical currents, to find their connection to magnetism, and eventually, for Maxwell to fully unify the two forces.

Thomas Young (1773–1829). One of the contenders for the title "Last Man to Know Everything," Thomas Young's knowledge was broad and far reaching. He studied medicine and the optics of the human eye and was one of the first scholars to translate the Rosetta Stone, which allowed us to read Egyptian hieroglyphs. He was born as the youngest of 10 children to a Quaker family in England and, by the age of 14, was said to be able to read 12 ancient languages!

For physicists, Young's most famous work was his double-slit experiment, in which he passed a beam of light through two narrow slits and observed a diffraction pattern on a screen on the other side. Even in modern physics, the results of this experiment are used as one of the strongest pieces of evidence in favor of the wave nature of light, though it would take until Maxwell's equations to demonstrate just what light is a wave of.

André-Marie Ampère (1775–1836). A French scientist and professor, Ampère was one of the first and most successful to expand on Oersted's connection between electric and magnetic forces. Ampère's name is now attached to the law describing the interaction of currents, reducing magnetism to the result of the motion of small charge carriers. Ampère's ultimate legacy to physics is overshadowed by that of Maxwell, but he was a pioneer in the study of electrodynamics, as Newton and Galileo were for mechanics (although he was not nearly so colorful a personality!). Although later scientists are given credit for more fully developing the theory, Ampère was one of the giants on whose shoulders they stood. Ampère spent the latter part of his career (after 1827) studying philosophy, which he considered "the only really important science."

Hans Christian Oersted (1777–1851). The name of this Danish physicist would likely be lost to obscurity were it not for happenstance. While preparing for a public lecture in 1820, Oersted noticed that a moving current caused a deviation of a nearby compass needle. After some intensive investigation, he published this discovery, which provided the impetus to other scientists (Ampère and Faraday foremost among them) to develop the mathematical and conceptual frameworks for understanding this phenomenon, culminating in the work of Maxwell.

Oersted's discovery would not have been so sensational except that it was in direct contradiction to the hypothesis of Coulomb, which had been taken as fact, that there categorically could be no interaction between electricity and magnetism. It took a direct repetition of the experiment, two months after the initial publication, in front of the

French Academy for that body to accept the data as something other than blatant falsification.

Johann Carl Friedrich Gauss (1777–1855). Gauss's brilliance was revealed early: The story is that by the age of 7, he amazed his teachers by almost instantly summing all the numbers between 1 and 100 (by the trick of realizing that he could rearrange the numbers into 50 pairs that each added to 101 [1 + 100] + [2 + 99]…).

Gauss published works in both mathematics and practical astronomy, collecting observations used to further refine the known orbits of planets for 30 years. His mathematical work was concerned with differential geometry, inspired by an early job in surveying. Perhaps his greatest contribution to physics, though, was his work on potential theory—an alternative means of representing forces felt by an object, or fields, resulting in the mathematics of Gauss's law, the first of Maxwell's equations (describing the relationship between an electric field and the charges that create it). Gauss's other accomplishments included constructing a primitive telegraph and estimating the position of the magnetic south pole of the Earth. After his friend Wilhelm Weber was forced to leave the University of Gottingen (where Gauss himself taught), Gauss became less involved in active research, preferring to follow the developments of younger mathematicians.

Michael Faraday (1791–1867). Born the son of a blacksmith, Faraday was a self-educated bookbinder who was hired as a laboratory assistant of Humphrey Davy in 1812, with his sole recommendation being a complete set of notes on Davy's own public lectures. Faraday's humble upbringing put his early career at odds with the "gentlemen's" society of early-19th-century physics, and his treatment at the hands of Davy (who blocked Faraday's admission to the Royal Society for six years) and others almost caused him to leave science altogether.

Having learned physics with no formal training in mathematics, Faraday's work was a model of tremendous mathematical intuition and incisive analogy, with relatively little formal mathematical development. This methodology put him at odds with the likes of Ampère and Maxwell, whose work focused on careful mathematical deliberation. Across a scientific career that spanned 40 years, Faraday is most well known for his development of the concept of the electromagnetic *field*, an intermediary in the interaction of two objects at a distance.

His work in the 1830s is one of the first conclusive discoveries of electromagnetic induction—the use of changing electric and magnetic fields to create currents—which he then harnessed to build the first dynamo, a device that converts mechanical energy into electrical current and vice versa and serves as the basis for modern electrical generators. Faraday is also credited with discovering the first connections between magnetism and light, which opened the way for Maxwell's later (and most important) work.

James Clerk Maxwell (1831–1879). A Scotsman, born without privilege or high social rank, Maxwell worked in the fields of mathematical physics and electricity and magnetism during the 1800s, when the scientific community was tackling this "exotic" subject with great vigor. Maxwell was especially intrigued by the discoveries of Michael Faraday (himself a man of humble beginnings), who had introduced the concept of force field as a physically relevant entity. Maxwell succeeded in mathematically describing *all* phenomena of electric and magnetic origin in a set of four relatively simple equations, now called *Maxwell's equations*. For the most part, these equations had been developed over the previous decades by others, but Maxwell organized and formalized them and added a key component, based not on experiment but his own aesthetic mathematical sense of symmetry, intimately and permanently unifying electricity and magnetism. Maxwell discovered that these equations lead to the phenomenon of "traveling electromagnetic radiation," moving at the speed of light, and with this, he realized the deep connection of electricity and magnetism to optics, as well.

Today, Maxwell's equations and the corresponding unification of forces are regarded as one of the grand highlights of human intellectual achievement. They form the basis of electrical engineering and modern optics and have survived the discoveries of modern physics in the 20th century essentially unscathed. They paved the way for the discovery of relativity (being fully relativistic equations, even though Maxwell hadn't appreciated that!) and form the classical underpinnings of quantum electrodynamics, the quantum theory of light. Studying Maxwell's ideas generally forms the second half of any standard college-level introductory physics course.

Heinrich Hertz (1857–1894). Hertz studied under some of the finest minds in German universities and obtained his Ph.D. in physics by 1880, at the age of 23. He experimentally demonstrated the existence of electromagnetic waves, as predicted by Maxwell's equations. He also discovered the photoelectric effect, though it would take Einstein to explain its origin. Hertz died at the age of 37 (of blood poisoning). Marconi followed up quickly on Hertz's experiments as a means to send signals, with the invention of the radio.

William Thomson, Lord Kelvin (1824–1907). William Thomson is, in many ways, the bridge between Faraday's intuitive brilliance and Maxwell's ultimate formulation of electrodynamics. Raised by his father, a widowed professor of mathematics, William was precocious in his scientific exploits. By the age of 15, he had already won a prize for an "Essay on the Figure of the Earth," and he began publishing papers (under a pseudonym) by age 16. By the time he was 22, he had earned a position as the chair of natural philosophy at the University of Glasgow, a position he held for 53 years. In 1845, Thomson began corresponding with Faraday, and the two men established a relationship of mutual respect. What Faraday approached intuitively, Thomson approached with mathematical models, including the mathematics developed by his lifelong friend George Stokes in studying heat flow.

Thomson's correspondence with James Maxwell led to the latter's tackling the problem of mathematically expressing Faraday's "lines of force," an effort that culminated in what we now call Maxwell's equations.

Bibliography

Essential Reading:

Hewitt, Paul. *Conceptual Physics.* Reading, MA: Addison Wesley, 2005. This is a textbook for a course perhaps a little more technically oriented than ours, but it's really wonderful. Hewitt is very accessible, with a strong focus on sense-making and understanding. Highly recommended to go along with this course if you want to push a little farther. Be aware: Trying to "read" a textbook like this one is a difficult task. You can't read it like a work of literature (much less like the daily paper or a novel)—it requires time for calculations, projects, and reflection.

Hobson, Art. *Physics: Concepts and Connections.* Englewood Cliffs, NJ: Prentice Hall, 2003.This is a textbook for a traditional course very much like ours. Aimed at the nonscientist (no algebra, minimal use of graphs and numbers), it's a good survey of the field. Hobson follows four themes: how we know science, post-Newtonian physics (which is not an emphasis of this course!), energy, and the social context of physics. Hobson also follows a quasi-historical path, with quite a bit of discussion about the nature of science and the context and significance of the big ideas in physics.

March, Robert H. *Physics for Poets*. New York: McGraw-Hill, 1996. This book is very much on the level and style of our course. This one is not a conventional physics text at all; it has a few equations but doesn't fuss with their manipulation. It is much more a historical overview of the big ideas and central characters of physics. A good companion to this course, although a bit brief if you become interested in any given individual and, indeed, a little superficial (focusing on the "ideal physicists" rather than troubling itself with historical complexities) if you have a historical bent, but a good start for getting into this material.

Pollock, Steven, and Ephraim Fischbach, *Thinkwell Physics I* (www.thinkwell.com). This is a multimedia video textbook, a collection of 10-minute "mini-lectures" by yours truly, covering much of classical mechanics, plus waves and oscillations. These lectures are designed to go along with a much more traditional physics course, but if you concentrate on the introductory lectures in each topic (rather than on the ones focused on calculating and problem-solving), they should complement the material in this course nicely. And if you decide you do want to delve a little farther into the mathematics on your own, *Thinkwell* will certainly be a useful guide.

Recommended Reading (referenced explicitly in this course):

Crease, Robert. *The Prism and the Pendulum.* New York: Random House, 2003. A lovely book, aimed very much at the audience for this course. His theme is that science and scientific experiments can be beautiful—not in some abstract way, not stretching the definition of the word, but meaning precisely what we always mean by *beauty*. Science and scientific experiments convey harmony, symmetry, and depth; they lead us to realizations about ourselves and the world; they change our outlook in positive ways; and they make us happy. Crease has picked 10 great experiments and explains them clearly and compellingly. Although the last few reach the realm of modern physics, this book is a nice complement to this course.

Cropper, William. *Great Physicists: The Life and Times of Leading Physicists from Galileo to Hawking*. New York: Oxford University Press, 2001. Short chapters on about 30 of the most influential physicists from Galileo to Hawking, with details on both the people and the physics they discovered. There are some equations, though the math is not a heavy emphasis, and they are often treated separately from the conceptual and historical discussions. Brief by its nature but well written and a nice mix of culture, significance, and physics itself. I learned a lot from this book!

Feynman, Richard P., Robert B. Leighton, and Matthew Sands. *The Feynman Lectures on Physics*. Reading, MA: Addison Wesley, 2005. I have to include this textbook, although "reading" it is essentially an impossible task for someone not already familiar with the basics of physics. Feynman sat down with the goal of presenting the great fundamental ideas of physics at an introductory college level; the result is this compilation of notes/text for his extraordinary freshman physics course at CalTech in the 1960s. He reformulated the traditional "canon" based on his own ingenious insights, creativity, and novel point of view. Once you've got some solid understanding of the basics of physics (even somewhat beyond where this course will take you), going back to this text will be a pleasure and a reward.

Gleick, James. *Isaac Newton.* New York: Vintage Books, 2004. An engaging biography of Newton that discusses the physics only qualitatively but sets a clear background of the context and culture in which Newton worked and the significance of his work. Detailed and full of insight into Newton's personality, this book paints a more complete picture than most biographies.

Gonick, Larry, and Art Huffman. *The Cartoon Guide to Physics*. London: Collins, 2005. I know that these *Cartoon Guides* may look superficial, but I'm a fan of this series. The coverage is solid, and the books are clever and fun to read. This book matches well with our course, and there's a nice mix of representations—I believe the cartoons do help make sense of the basic ideas of classical physics.

Lightman, Alan. *Great Ideas in Physics*. New York: McGraw-Hill, 2000. Lightman zooms in on only four "great ideas" (two from classical physics, energy conservation and the second law of thermodynamics). His perspective melds physics, philosophy, and art, although he focuses on the physics, walking you through a little bit of the mathematics to get a taste for the role of math in understanding. A little limited in scope, but useful if you would like to begin the trip from conceptual physics to mathematical physics without taxing your math skills (you need to be comfortable with ratios and basic algebra). The questions for reflection at the end of the book are particularly good.

Further Recommended Reading:

Asimov, Isaac. *The History of Physics*. New York: Walker & Co., 1984. Asimov has produced a readable, comprehensive history of physics, although it's not so much history as it is details and concepts mixed in with history, biography, and philosophy of science.

Christianson, Gale. *Isaac Newton* (Lives and Legacies Series). New York: Oxford University Press, 2005. A brief, simple introduction to Newton and his physics. Although it's a little less nuanced than Gleick's biography (listed above), I nevertheless enjoyed this selection.

Cohen, I. Bernard. *The Birth of a New Physics*, rev. ed. New York: W.W. Norton, 1991. Good historical coverage of the physics of the early scientific revolution, particularly in the 16th and 17th centuries.

de Campos Valadares, E. *Physics, Fun, and Beyond: Electrifying Projects and Inventions from Recycled and Low-Cost Materials*. Englewood Cliffs, NJ: Prentice Hall, 2006. This is a collection of simple, "at-home" experiments and projects, spanning much of classical experimental physics, suitable for science fair ideas, family projects and gifts, teaching/outreach, or just plain interesting hobby activities. Great for those who prefer to learn by doing, although the author also takes care to explain the physics behind each of the projects.

Ehrlich, Robert. *Turning the World Inside Out and 174 Other Simple Physics Demonstrations*. Princeton, NJ: Princeton University Press, 1990. Another collection of physics activities and demonstrations, this one is aimed a little more at a teacher, but it provides inspiring ideas for anyone interested in watching physics in action in clear, simplified ways. Each project has detailed construction instructions and physics explanations, and almost all the projects are quite simple, requiring relatively little in the way of expense or equipment.

Epstein, Lewis. *Thinking Physics: Understandable Practical Reality*. San Francisco: Insight Press, 2002. A wonderful collection of cartoon-based "thinker" puzzles, designed to see if you have a strong conceptual understanding of many of the topics of classical physics, such as how tides work or why steel ships float. These questions are often designed at the level of an introductory college course (some of them involve sense-making of the mathematics in a traditional physics class), but by and large, there is no calculation of any kind required for these questions, just clear thinking about the underlying principles of physics. Epstein is great at talking through the wrong answers to help you "think about your own thinking."

Ferguson, Kitty. *Tycho and Kepler: The Unlikely Partnership That Forever Changed Our Understanding of the Heavens*. New York: Walker & Co., 2002. A good biography of these two remarkable historical figures that includes some of the essential physical ideas, highlighting the brilliance of Kepler's achievements.

Feynman, Richard. *The Pleasure of Finding Things Out*. New York: Basic Books, 2005.

———. *Six Easy Pieces: Essential Physics Explained by Its Most Brilliant Teacher*. New York: Basic Books, 2005.

———. *The Character of Physics Law*. New York: Modern Library, 1994.

———. *The Meaning of It All*. New York: Basic Books, 1998.

Richard Feynman is one of the great 20th-century physicists, and his perspectives on the nature of science are unparalleled. *The Pleasure of Finding Things Out* is a collection of Feynman's essays on a number of topics, offering nontechnical but delightful insights into how science is done. *Six Easy Pieces* is a collection of the least technical chapters from the *Feynman Lectures*, in which he introduces big topics of (mostly, with one or two exceptions) classical physics ideas. *The Character of Physical Law* is in a similar style, focusing on some central topics of physics and talking both about the details and the "meta" issues, the nature and consequences of science. *The Meaning of It All* drifts farther from the physics and into issues of the connections among science, religion, and politics. There are many other books by (and about) Feynman, all of which are highly recommended, although some go beyond the "classical" focus of our course.

Gamow, George. *The Great Physicists, from Galileo to Einstein.* New York: Dover Publications, 1988. George Gamow, inventor of the Big Bang theory, is a skilled author for non-physicists. Gamow's books are gems, inspiring, and suitable even for young adults. This "biography" of physics covers much of the classical physics topics we've focused on, ending with some discussion of modern ideas.

Gribbin, John. *The Scientists: A History of Science Told through the Lives of Its Greatest Inventors.* New York: Random House, 2004. Five hundred years of science in 672 pages—this is a comprehensive book, written compellingly, with anecdotes and stories. Gribbin organizes and connects the characters and developments. A reference (you can wander from spot to spot in the book if you want) that also makes for a compelling read, albeit a little heavy going.

Heilbron, John L. *The Oxford Guide to the History of Physics and Astronomy.* New York: Oxford University Press, 2005. An encyclopedic collection (quite literally) of information about the personalities and the physics; very complete, informative, and surprisingly interesting just to read.

Holton, Gerald, and Stephen Brush. *Physics, the Human Adventure: From Copernicus to Einstein and Beyond.* New Brunswick, NJ: Rutgers University Press, 2001. Teaching physics with an accurate historical and philosophical perspective, with more history than March's *Physics for Poets*.

Jargodzki, C., and F. Potter. *Mad About Physics: Braintwisters, Paradoxes, and Curiosities.* New York: Wiley, 2001. Another collection of physics puzzles; maybe just a little less "physics-serious" than Walker's *Flying Circus* (see below), it provides briefer explanations to a broader assortment of puzzles. This entertaining book takes advantage of paradoxes as a teaching tool and includes wonderful quotes.

Jungnickel, Christa, and Russell McCormmach. *Intellectual Mastery of Nature: Theoretical Physics from Ohm to Einstein*, 2 vols. Chicago: University of Chicago Press, 1990. This book focuses on the emergence of theoretical physics as a discipline, mostly in Germany and Austria, between 1850 and 1925, offering a largely biographical development and context. A scholarly work; again, a little heavy going but particularly appropriate for the electricity and magnetism section of this course.

Kakalos, James. *The Physics of Superheroes.* New York: Gotham, 2005. I may have a soft spot for whimsical physical texts, but this one strikes me as very successful at teaching the basic principles of classical physics in the context of comic-book superheroes. The comics provide a framing for Kakalos to teach the basic principles of physics in an engaging way.

Kuhn, Thomas. *The Structure of Scientific Revolutions*. Chicago: University of Chicago Press, 1996. Kuhn is a philosopher of science who popularized the notion of paradigm shifts. This book discusses the nature of the evolution and progress of scientific ideas. Kuhn argues that, for the most part, scientific progress is incremental and exists within a scientific (and sociological) framework; only rarely are "revolutions" possible. Some traditional physicists disagree with Kuhn's arguments regarding the extent to which scientific progress is socially constructed, but the work is interesting, challenging, and influential.

MacAulay, David. *The New Way Things Work.* Boston, MA: Houghton Mifflin, 1998. A cartoon-based book, whimsical and amusing, it takes an engineering and physics approach to examine devices and ask how they work, incorporating physics concepts in a meaningful way. Aimed at children, but child-like adults (like me) can appreciate this book.

Mahon, Basil. *The Man Who Changed Everything: The Life of James Clerk Maxwell.* New York: John Wiley & Sons, 2004. A brief and readable biography of Maxwell's life and science. It doesn't get so much into the physics but offers good insights into Maxwell as a human being and scientist!

Nye, Mary Jo. *Before Big Science: The Pursuit of Modern Chemistry and Physics, 1800–1940*. Cambridge, MA: Harvard University Press, 1999. A thorough study of the two sciences together, emphasizing the social and historical context.

Purrington, R. D. *Physics in the Nineteenth Century*. New Brunswick, NJ: Rutgers University Press, 1997. A little heavy going, this is pure historical analysis but covers all the major players in the physics of the 1800s, with an emphasis on the development of ideas leading to the coming revolutions of the 20th century.

Shamos, Morris, ed. *Great Experiments in Physics: Firsthand Accounts from Galileo to Einstein*. New York: Dover Publications, 1987. A collection of 25 key experiments (including many discussed in this course), introduced and explained, then followed with annotated original works. A unique book; it's great (and surprisingly rare) to read the originals.

Spielberg, N., and B. Anderson. *Seven Ideas That Shook the Universe*. New York: Wiley, 1995. A comprehensive text, covering the ideas of this course and following a conceptual framework. Aimed at the non-physicist. A highly readable text.

Vollmann, William. *Uncentering the Earth: Copernicus and the Revolutions of the Heavenly Spheres*. New York: W.W. Norton, 2006. A book (written by a nonscientist) that explores how Copernicus could have come up with the heliocentric hypothesis and convinced himself of its correctness. Not always the easiest read but a fascinating story that gets at the core of the start of the scientific revolution and the nature of scientific reasoning.

Von Baeyer, Hans Christian. *Warmth Disperses and Time Passes: The History of Heat*. New York: Modern Library, 1999. A well-written historical and physical treatment of thermodynamics. A good introduction to the story of thermodynamics, which alas, we barely have time to even introduce in this course.

Walker, Jearl. *The Flying Circus of Physics with Answers*. New York: Wiley, 1977. A collection of "puzzles," all curious, real-world phenomena for you to think about, that demand physical explanation: Why does chalk squeak? How does a one-way mirror work? What's the "green flash" at sunset? Why wasn't Ben Franklin killed when he flew his kite in a lightning storm? These questions are a lot of fun to explore! The puzzles are organized around broad themes of classical physics (such as mechanics, optics, acoustics, thermodynamics, and so on).

Standard Introductory Physics Textbooks (a selection):

A number of textbooks are used in introductory college-level physics courses. These are not generally designed as "standalone" reading but are meant to be used with the guidance and support of an instructor. If you decide to buy one to try out on your own, be advised that these are not "evening reading material." There are so many, I list below only a few of my personal favorites (the one I authored, *Thinkwell Physics*, was listed with the essential texts, above). Many others are in use in college courses around the world; this is a very abbreviated list!

Bloomfield, Louis. *How Things Work: The Physics of Everyday Life.* New York: Wiley, 2005. An unconventional introduction to physics, aimed at nonscientists who want to learn the basic principles of physics and the applications to everyday life. Rather than organizing the text around physics concepts, the author focuses each chapter on a technological or physical application (generally both common and interesting!). This approach develops the physical principles in a deeply motivating way. Of all the texts listed in this section, Bloomfield's is likely to be the most accessible to the interested layperson, but even so, it remains a textbook that would probably best be used in the framework of a course with an instructor.

Chabay, Ruth, and Bruce Sherwood. *Matter and Interactions*. New York: Wiley, 2003. Most of the standard introductory texts follow pretty much the same pattern, teaching the same classical physics topics in roughly the same order (perhaps adding modern physics in the end) and focusing on the same mathematical skills. This text offers a fresh approach. Sherwood and Chabay are part of the physics education research community and treat introductory physics from a completely modern perspective. Relativity and the atomic model are involved right from the start, and the separation between classical and modern physics is purposefully blurred. The authors emphasize modeling systems throughout. If you want to learn physics with the intent of becoming a physicist, this would be an excellent first textbook to use, but again, the level of mathematics and sophistication required is fairly high; this is certainly not "light reading."

Giancoli, Douglas. *Principles with Applications*. Englewood Cliffs, NJ: Prentice Hall, 2004. This is a fairly traditional and popular introductory textbook, designed specifically for an algebra-based course. Many of the applications and examples in the book are tailored to students who are less likely to be physicists or engineers but might be interested in medicine, biology, or architecture.

Halliday, David, Robert Resnick, and Jearl Walker, *Fundamentals of Physics* (New York: Wiley, 2004), or perhaps, Karen Cummings, Priscilla Laws, Edward Redish, and Patrick Cooney, *Understanding Physics* (New York: Wiley, 2004). Halliday, et al., has been one of the standard texts at many schools for many years. As you move up to more recent editions, there is a stronger focus on conceptual understanding. The *Understanding Physics* book is basically a new, updated version, redesigned to incorporate physics education research results, but it is nevertheless still a dense, heavy, mathematically centered introductory text. It remains one of my favorites for teaching calculus-based physics and engineering courses.

Knight, Randall. *Physics for Scientists and Engineers: A Strategic Approach*. Reading, MA: Pearson/Addison-Wesley, 2003. Similar in content to Halliday, Resnick, and Walker, above. Knight has also taken a stab at rewriting the conventional introductory calculus-based textbook with physics education research results in mind. That means using research on common student learning difficulties, incorporating alternative representations and metaphors, and including problems and questions designed through iterative research studies.

Moore, Thomas. *Six Ideas That Shaped Physics*. New York: McGraw-Hill, 2003. This is another modern, nonstandard approach to the introductory text. Breaking the subject into six fundamental "big ideas" (such as conservation laws, reference frame–independence of physics, universal laws, and so on), Moore leads the student to apply basic principles to solve realistic physical problems, rather than following a more traditional, plug-'n-chug, formula-centric approach.

Internet Resources:

The Web has an overwhelming supply of resources regarding introductory physics (*some* of which are even accurate and useful)! The task of selecting just a few sites is difficult (and the situation will likely evolve so quickly as to limit the usefulness of this list), but below are a few Web sites that I believe are definitely worth investigating.

http://phet.colorado.edu. This is the simulation site referred to throughout this course, developed by the Physics Education Research group at the University of Colorado.

http://natsim.net/en.html. This site contains links to other physics simulation collections. Although the phet sims (listed above) are very helpful, they cover only a narrow range of topics. This page will take you to sites with hundreds of applets. In addition, you might want to visit sites mentioned explicitly in the lecture notes:

- www.cecm.sfu.ca/~scharein/astro
- www.walter-fendt.de/ph11e
- http://physics.bu.edu/~duffy/semester1

http://howthingswork.virginia.edu/. Louis Bloomfield (whose introductory textbook for nonscientists, *How Things Work*, is also on my recommended list) has created a high-quality frequently-asked-questions page for explanations about the physics of everyday life. If you have a question about a device or phenomenon, there's a pretty good chance you will be able to find an answer on this page.

http://www.merlot.org/merlot/materials.htm?category=2737. The Merlot Web site (www.merlot.org) is a national resource for academics in a variety of fields to compile learning materials. The link above takes you specifically to a collection of peer-reviewed resources for classical mechanics. (Moving up a level will allow you to explore more of physics, including electricity and magnetism and modern physics)

www.aip.org/history/syllabi/books.htm. The AIP is the American Institute of Physics. This is the institute's "bibliography" page, with many highly recommended books. (Some of them I have listed above, but I've tried to keep my bibliography distinct. AIP's selection is very good!)

www.aip.org/history/gap/. Another AIP page, this one has links to the works of some great American physicists (including original papers, with explanations), including Franklin, Gibbs, and many others.

www.physlink.com/Education/History.cfm. A collection of links to other sites, with history and timelines. Also many links to science museums.

www-gap.dcs.st-and.ac.uk/~history/index.html. This page calls itself the History of Mathematics Archives, but it's very thorough, and its compilers apparently consider most physicists to be mathematicians. The biographies are interesting and comprehensive without being overwhelming. This is my favorite site for a "quick read" about some historical figure I'm interested in.

www.upscale.utoronto.ca/PVB/PVB.html. This is the University of Toronto's *Physics Virtual Bookshelf*. The staff at the university has put together an impressive collection of links and articles. A nice place to start digging deeper into the history and content of physics.

http://en.wikipedia.org/wiki/Physics. Wikipedia is a collective, informal, Web-based encyclopedia. This site is frequently helpful, and I use it all the time (not just for physics!). But beware: It is the nature of Wikipedia that there can, on occasion, be mistakes or even sheer nonsense here. These articles are submitted by individuals without "authorization"; this is not the usual method of scientific peer review by any stretch of the imagination. If you learn something here, follow up to make sure that it's accurate and reliable. Nevertheless, Wikipedia is often my first stop when I'm looking up something new.

www.physics.org/. From the Institute of Physics, many links and interactive sites for history and the "physics of everyday life."

http://physicsweb.org/bestof/history. Another compilation of historical information and links, this one put together by the Institute of Physics (IoP).

www.hssonline.org/teach_res/essays/mf_essays.html. A recommended bibliography from the History of Science Society. Once again, many good books here, organized in a variety of categories (social, historical, bibliographic). An excellent resource for delving further into the history of classical physics!

http://galileo.rice.edu/. A comprehensive Web site about Galileo.

www.tychobrahe.com/eng_tychobrahe/index.html. A comprehensive Web site about Brahe.

www.clerkmaxwellfoundation.org/. A comprehensive Web site about Maxwell.

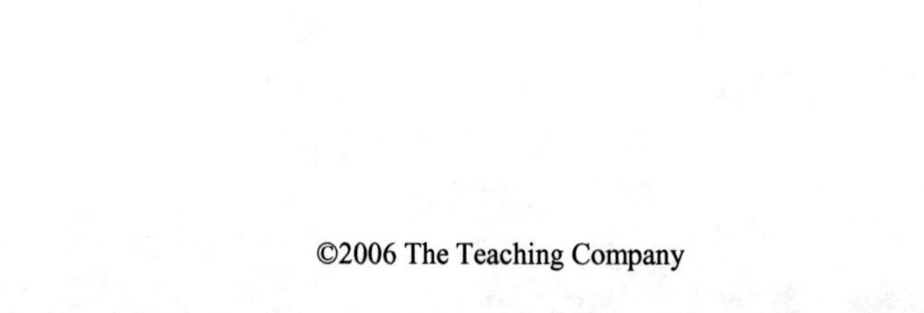